The Collector's World of Inkwells

Jean & Franklin Hunting

4880 Lower Valley Road, Atglen, PA 19310 USA

Dedication

We dedicate this book to the people who so graciously opened their homes to us so that we could photograph their wonderful inkwells. We enjoyed the many hours we spent with you and want you to know how much we appreciate your assistance. Your willingness to share your inkwells with others is certainly commendable.

James L. Adams	Trudy and Lee Geer
Gary Bahr*	Jerry Hentzler
Rhonda L. Bartlett	Lou Krumpholz
Steven J. Bohm	Merlin J. Nomann
Marcia and Dell Breon	Dan Prewitt
Margaret T. Buttrey	Dorothy and Robert Soares
Johnny M. Crabb	Judee and Kip Vogt
Diane and Roy Farriester	Douglas Warcup

*Gary Bahr is well known for the many interesting and informative articles he has written about inkwells and inkstands that appear in *The Stained Finger*, a newsletter published by the Society of Inkwell Collectors.

Library of Congress Cataloging-in-Publication Data

Hunting, Jean.
 Collector's world of inkwells / Jean & Frank Hunting.
   p. cm.
 Includes bibliographical references.
 ISBN 0-7643-1102-6
   1. Inkwells--Collectors and collecting--Catalogs.
   I. Hunting, Franklin. II. Title.

NK6035 .H86 2000
681'.6--dc21
                                          99-059828

Copyright © 2000 by Jean & Franklin Hunting

All rights reserved. No part of this work may be reproduced or used in any form or by any means—graphic, electronic, or mechanical, including photocopying or information storage and retrieval systems—without written permission from the copyright holder.
"Schiffer," "Schiffer Publishing Ltd. & Design," and the "Design of pen and ink well" are registered trademarks of Schiffer Publishing Ltd.

Book Design by Anne Davidsen
Type set in Lydian /Aldine721

ISBN: 0-7643-1102-6

Printed in China
1 2 3 4

Published by Schiffer Publishing Ltd.
4880 Lower Valley Road
Atglen, PA 19310
Phone: (610) 593-1777; Fax: (610) 593-2002
E-mail: Schifferbk@aol.com
Please visit our web site catalog at **www.schifferbooks.com**
We are always looking for people to write books on new and related subjects. If you have an idea for a book, please contact us at the above address.

This book may be purchased from the publisher.
Include $3.95 for shipping.
Please try your bookstore first.
You may write for a free catalog.

In Europe, Schiffer books are distributed by:
Bushwood Books
6 Marksbury Ave.
Kew Gardens
Surrey TW9 4JF England
Phone: 44 (0)208 392-8585; Fax: 44 (0)208 392-9876
E-mail: Bushwd@aol.com
Free postage in the UK. Europe: air mail at cost.
Try your bookstore first.

Contents

Acknowledgments 4

History of the Society of Inkwell Collectors 5

Introduction 6

Price Guide 8

Ink 8

Pounce Pots & Sand Castors 9

Quills & Pens 10

Inkwells & Inkstands 12

Travelers or Pocket Inkwells 14

U. S. Design Patents & Their Time Frames 15

U. S. Invention Patents & Their Time Frames 15

A Gallery of Inkwells 16
      Traveler's Inkwells 275

References 286

Index 287

Acknowledgments

A Special Acknowledgment

We wish to express our gratitude to *Bernard J. Barkoff* of London, England, who was kind enough to review and edit this book. We met Bernard at the Society of Inkwell Collectors Convention that was held in Houston, Texas, in June of 1999. We were most impressed with his vast knowledge of inkwells and were very pleased when he offered to review the book for us.

Bernard has been a dealer and collector of inkwells for over thirty years. He is the author of the book *Writing Antiques*, by Shire Albums, London, England. For the past ten years he has been a licensed appraiser of inkwells. He is also an excellent auctioneer, a member of the Society of Inkwell Collectors, and considered by the Society to be an expert in the field. He travels to the United States often to do appraisals for inkwell collectors.

Other Notable Contributors

We wish to express our appreciation to our friends in the antiques business. They exhibited a great deal of patience and good humor while we disrupted their routine at the shows and their shops with our pesky picture taking. They were most generous with their time and knowledge. We truly could not have put this book together without you. We thank you all!

Barbara and Steve Aaronson, Victorian Lady Antiques, Northridge, California
Bob Albert, Landmark Antiques, Dana Point, California
Brian Barton and Susan Bondesen, Exeter Antiques, Exeter, California
Bondé A. Bliven, Bondé Antiques, Carmel, California
Cyril Boyce, Ashford Antiques, San Francisco, California
Charlene N. Cranfill, Antiquetessen, Wildomar, California
Patricia Gaualdá, Peace Sign Antiques, Buenos Aires, Argentina
Beverly Graham, Graham & Co. Antiques, Clovis, California
Frances and Roy Hintergardt, Homestead Antiques, Sanger, California
Shari Keeler, Shari's Antiques, Portland Oregon
Dr. Sheppard B. Kominars, The Good Sheppard Antiques, San Francisco, California
Susan Lowry Antiques and Interiors, Chesterfield's Antiques, Fresno, California
Carolyn Meeker, L. C. Collectibles, Kingsburg, California
Judy and Ray Mesick, Antique Arcade, Redwood City, California
Pat Mizrahi, Mizrahi Antiques, Sherman Oaks, California
Chuck Morgenstern, The Woodchuck, San Francisco, California
Eve Naslund, Eve Naslund Antiques, El Cajon, California
Barbara Nelson, Barbara Nelson Antiques, Laguna Beach, California
Muriel Peterson, Muriel Peterson Antiques, Richmond, Washington
Dorothy and Robert Soares, Soares Antiques, Stockton, California
Susan and Bill Tanner, Susan Tanner Antiques, Yorba Linda, California
Sam Torcaso, Chesterfield's Antiques, Fresno, California
Dan Wade, Collector's Mall, Oakhurst, California
Joan and John Walter, Walter's Antiques, San Jose, California

History of the Society of Inkwell Collectors

by Vincent McGraw

In 1965, while attending an auction, I purchased my first inkwell in or around Crawfordsville, Indiana. It was then, and still is, my favorite inkwell.

My collection continued to grow through flea markets, antique shows, garage sales, gifts, and wherever. My interest also continued to grow. I visited libraries and bookstores. I wrote to anywhere and everywhere I could to find information. As most of you know, most of my efforts were in vain. The little information that was available was all the same, and very sketchy. Then along came books by Covil, Rivera, Walter, and eventually McGraw and Goodman. Badders' book has now joined the field. All are different, and all show the wide variety of inkwells that are out there.

My book, published in 1972, generated a lot of correspondence, a lot of friends, a lot of questions, and a lot of answers. Most collectors are eager to learn more about every piece in their collection, check prices, and share anecdotes about collecting with other enthusiasts. In the past, this sort of camaraderie has been difficult to attain for the far-flung inkwell collectors.

Because of this need, the Society of Inkwell Collectors was formed in 1981. Its purpose is to introduce inkwell collectors from around the world, help them keep in touch, and share information about inkwell collecting. This is accomplished through our quarterly newsletter, *The Stained Finger*, and through our conventions.

Membership now includes 700 collectors in countries as diverse as Australia, Belgium, Canada, England, France, Hong Kong, Italy, Japan, Spain, and Germany. We have members from all 50 United States and the District of Columbia. We are sharing information, and making new friends in the process.

We have broadened our interests over the years, going beyond the fascinating subject of inkwells to cover most of the history of writing accessories. We encourage the study of these items as well as their conservation for future generations. But most of all, we encourage a love of writing, both through a study of its history and through appreciation of its importance to us today.

>Vince McGraw
>Society of Inkwell Collectors
>5136 Thomas Avenue South
>Minneapolis, MN 55410

Inkwell collectors owe Vince McGraw a debt of gratitude. He is the founder of the Society of Inkwell Collectors. The Society puts out a quarterly newsletter, *The Stained Finger*. There are excellent articles to peruse, and members share knowledge through a question and answer section. Anyone interested in inkwells would do well to join. Another advantage, those hard-to-find glass inserts are available through the Society.

Introduction

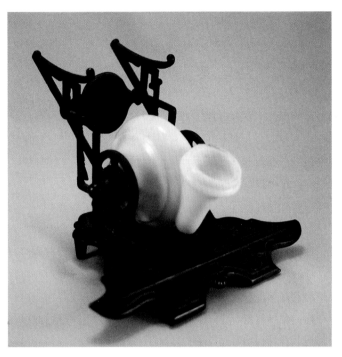

Our interest in inkwells was sparked by one that we inherited from the family—an inkstand with a revolving milk glass well in the shape of a snail. It is attached to an iron frame and rotates up against a round stopper disc to close. The design and the fact that it was made in the United States circa 1880 intrigued us. Although we have collected many other inkwells over the years, this one remains our favorite.

When we first started collecting inkwells, there was scant information on the subject. Now there are several fine books available. We learned, sometimes the hard way, what we should look for before making a purchase. Unless an item is hallmarked, has a patent date, or is otherwise identified, it is impossible to be certain when and where it was made. However, there are clues to be found. We examine the top, bottom, sides, edges, and under the lid for patent dates, registry numbers, abbreviations for "patented" or "patent pending," any makers' marks, and/or a manufacturer's name. English sterling silver has hallmarks with a date letter and the initials of the maker, which makes identification fairly simple. Also, a few companies in the United States put date letters on some of their wares—Tiffany and Gorham come to mind. All are valuable clues that help identify and date a piece.

Influences in Inkwell Design

Most of the inkwells found in the United States today were made during the middle of the nineteenth century through the 1930s. Three style periods of design greatly influenced the manufacture of a vast array of items, including inkwells: the Arts and Crafts Movement (1861-1920), Art Nouveau (1890-1910), and Art Deco (1920s-1930s).

The Arts and Crafts Movement of the last half of the nineteenth century came about as a rebellion against mass-produced articles that were manufactured during the Industrial Revolution. In England, in 1861, William Morris founded a firm of interior decorators and manufacturers devoted to recapturing the quality of workmanship and the clean lines that were prevalent in previous years. He produced hand-crafted metalwork, furniture, textiles, books, et cetera. Many artisans joined the movement and hand-crafted wares began appearing worldwide. The Arts & Crafts style was popular in the United States from about 1895 to 1920.

Art Nouveau, new art, had its beginnings during the Arts & Crafts movement in England. Some distinguishing characteristics of the Art Noveau style are lovely, undulating, asymmetrical lines. For example, a woman's sinuous form with long flowing hair, the tendrils of a meandering vine, and flower buds and stems intertwined in a rhythmic embrace are some common images featured in art from Art Noveau period. This style flourished in the United States from about 1890 to 1910.

The Art Deco, or Style Modern, movement began in 1910. However, it was not until it appeared at the Exposition Internationale des Arts Decoratifs ét Industriels Modernes in Paris in 1925 that it acquired the descriptive name Art Deco. Modern, sleek looking lines, sharp angles, and geometric figures characterize the Art Deco style of design. The streamline looks of Art Deco are vastly different from the romantic flowing lines of Art Nouveau. This art style was popular in the United States during the 1930s.

Before Purchase

Before purchasing an inkwell, there are a few cautions to be taken. Condition is of utmost importance. A crack or chip will reduce the value significantly. Even iron will break when dropped, so look for broken edges on these items. A missing lid is almost impossible to replace. Over the years we have purchased several inkwells without lids, and we were sure we could find suitable replacements—we are still looking. Some inkstands require a separate liner, or insert, to hold the ink. A good many of these have been lost or broken. Inserts were made of glass, porcelain, pottery, or metal. We find many lovely inkstands marred by a gaping hole where an insert should be. An inkwell without the means to hold ink is like a car without a motor—it is not functional. An original insert is most desirable; however, a modern replacement will affect the value minimally.

Reproductions of desirable, but scarce items have been made since time began. A nineteenth-century reproduction of an eighteenth-century inkwell should not be scorned; it is still an antique. Unfortunately, today we have new and reproduced inkwells being sold as original antiques. Scent bottles are often mistaken for inkwells. So give a little extra thought to that pretty tall inkwell; it just might be a perfume bottle. Fakes are another thing to be aware of. We have seen inkwells converted from cigarette lighters. Believe it or not, we have even seen very good-looking inkwells made out of old doorknobs!

The ability to spot fakes and reproductions is very important. The best way to acquire expertise is to handle as many inkwells as possible, talk to knowledgeable dealers, and exchange ideas with fellow collectors. The Society of Inkwell Collectors offers a good forum.

Organization

We have measured the inkwells in this book as close as possible, stated where we believe they were made, and given an approximate date of manufacture. We have included a table of design and invention patents for those interested in further research. The patent date is not always the date of manufacture. Some inkwells were made over a long period of time or years after the patent was granted, but these dates are an important reference point. Only items with date letters such as those used in England represent the exact year of manufacture.

We have enjoyed putting this book together, it has indeed been a labor of love. Perhaps one of you will write the next inkwell book. We will be looking forward to it. There is so much yet to be learned.

Price Guide

It should be remembered that a price guide is just that, a guide. It is not intended to set prices or to be the final assessment of the value of an item. It is merely to give a broad range of the retail prices that can be reasonably found in the continental United States at the present time. The value of an item depends on many variables: rarity, beauty, size, condition, location, and, above all, inflation. A good example of how far inflation has brought us is an inkstand listed in this book with a value of $200-350 was advertised in a 1897 Sears and Roebuck Catalogue for 45 cents!!! When factoring for inflation, it must be remembered that salaries have risen as well as prices. In recent years inkwells and inkstands have increased in value dramatically.

Auctions can be useful as a pricing guide but caution must be exercised. Auction houses publish a list of merchandise to be auctioned off and give an appraisal of each piece; seldom do the expected prices match the final selling price. When two or more people desire the same object, the bidding can escalate far above the appraisal, or on a slow day bargain prices may be had. A department store has to set its prices in alignment with its competitors or the customer will go elsewhere because the same new merchandise is widely available. With antiques and collectibles that are no longer being made, it is not so simple. You cannot always find another identical item, and if you pass up a wonderful inkwell just because the price guide has it listed lower than the asking price, you may be making a mistake.

We arrived at these estimated prices using our own years of experience in the antiques business, checking current prices at antiques shows and shops, and consulting with many dealers and collectors across the United States. In addition, we had the expert advice of Bernard J. Barkoff, an internationally recognized appraiser and well-respected member of the Society of Inkwell Collectors.

In the long run the fluctuating market sets the prices. Due to the many variables that effect prices, items may be found selling for more or less than this estimate and neither the authors nor the publishers assume any responsibility for any losses that may be incurred by using this guide.

Ink

Ancient Egypt and China are believed to have used a form of ink as early as 2500 BC. Brown sepia, the liquid secretion from a gland of the cuttlefish, is thought to be the oldest known writing fluid. Around 1200 BC, China was making "India Ink," a mixture produced from carbon black stabilized with various substances: shellac in borax, gelatin, glue, gum Arabic, and dextrin. Colored solutions were made with dyes and pigments from berries, plants, soot, charcoal, and even animal blood. The ink was used in liquid form or molded and dried in blocks. The dried ink was rubbed off on an ink stone and saliva was added to make it liquid. Early ink was subject to molding, so it was necessary to mix it fresh on a daily basis.

Gall, a nut-like swelling left on an oak tree by a female wasp laying her eggs, was the basis for an improved ink. The gall nut was ground, soaked in water to extract tannic acid, strained, and mixed with copperas (hydrated ferrous sulfate) and gum Arabic. Gall ink was good when used with a quill; however, years later when steel pen nibs were introduced, its corrosive quality was ruinous. By 1860, advertisements began to appear touting inks that would not corrode steel pens.

During the eighteenth and nineteenth century, ink was made by a chemist or mixed in apothecary shops and sold in various size bottles. The liquid ink could be used directly from the small bottles in which it was purchased or transferred from a large bottle to an inkwell. Powdered ink was also available. Packages of dried ink were convenient for traveling, made for easy storage, and could be mixed with water as needed.

Pounce Pots & Sand Castors

Pounce is a fine powder made from pulverized sandarac, cuttlefish bone, or gum powder. It was spread over coarse paper to condition the surface for writing. Pounce was also used as a blotting agent, to dry wet ink and keep it from spreading or smearing. It was important to dry ink on a newly penned letter before folding or the writing would become blurred and the fingers stained with ink. Drying was sometimes achieved by spreading the wet paper in the open air, but a dusting of pounce was much quicker. Writing errors could be scraped off using a dual-purpose knife. One end has a sharp blade for trimming the quill point; the other end has a half-moon shaped blade ink eraser. The rough area left by the erasure was coated with pounce and rubbed smooth so a correction could be made. For centuries a pounce pot was incorporated into inkstands; it was unattached for easy removal and use. During the eighteenth and nineteenth century, individual pounce pots and sand casters were produced as a separate item also, usually matching an inkwell and other items in a desk set.

By the latter part of the eighteenth century, writing paper was greatly improved. Very absorbent, powdered sand had almost replaced pounce as a drying agent. Although other colors were available, black sand made from a form of mica was the most commonly used. A few inkstands had both a pounce pot and a sander in their design. Sand or pounce casters have small pierced holes in the top, similar to salt shakers of today. All have a recessed or concave top so any excess material can be recycled for future use. The finished document was rolled into a funnel and the unused portion was poured back in the top of the sander and sifted down through the perforated holes. Some sanders have a screw on top that opens for filling. They were made in materials such as wood, tin, pewter, glass, porcelain, pottery, silver, brass, or bronze. Sanders were made well into the latter half of the nineteenth century; their use declined as blotters made from porous paper became available.

Quills & Pens

Since the early cave dweller first discovered he could communicate with his fellow man by painting characters on the cave wall with charcoal, or scratching messages in the dirt with a stick showing the location of game, the pursuit of better methods of communication with the written word has been an on-going project. In ancient China, in about 1000 BC, a brush made of reeds was used for calligraphy. Early Egyptians used thick, sharpened reeds or bamboo to write hieroglyphics on papyrus. A pictographic type of cuneiform was used extensively in the Middle East; a sharp-edged stylus was pressed into soft clay tablets, thus forming wedge-shaped characters.

A quill pen, a feather with a split tip cut at an angle, has been used to transfer liquid dye to paper by capillary action for hundreds of years. Quill pens have been made from the outer wing feathers of turkeys, hawks, owls, pheasants, pelicans, peacocks, and even crows. Swan feathers were considered superior, but they were more costly. The 1908 Sears and Roebuck Catalogue advertised crow quill pens that were "fine and stiff .48 cents a dozen." Geese, being domesticated fowl, were the most common source of feathers. The feathers were collected during molting seasons, with each wing producing about five quills. Left-handed people favored the natural curvature of the right wing feather, while the right-handed favored the left wing feathers. Crow feathers provided a more delicate line and were often preferred by women. Most people dried and cut their own quills; however, they were made commercially as well. The feathers were treated with heat, dipped in hot sand to remove a thin skin from the shafts, and then cleaned in a solution of boiling alum, before being cut into pens. Over time different styles of writing dictated the way the point of the quill was cut. Some were made to form broad strokes, while others were cut to produce a fine script. Quills required frequent sharpening to keep an edge and over time were whittled down in size—so small they were barely useable. A special knife with a finely honed steel blade was used for this purpose. To preserve the quill it was necessary to store it in an upright position when not in use. A quill holder with a small amount of water, or lead shot, in the bottom protected the point.

The history of the United States, the Bill of Rights, the Declaration of Independence, and the Constitution were all written and signed by our founding fathers with quill pens dipped in ink. If you should visit George Washington's home at Mount Vernon, Thomas Jefferson's home at Monticello, Independence Hall in Philadelphia, or historic Williamsburg, Virginia, you will see the type of pewter inkwells and quill pens used in colonial days. Quill pens were used from about the sixth century to the middle of the nineteenth century.

Several attempts were made to invent a better writing instrument. Horn, tortoiseshell, and several different gemstones were tested, but steel proved to be the most effective. In 1828, John Mitchell of Birmingham, England, invented a machine to make steel pen nibs that could be fitted into a separate handle. Several small nibs could be kept on hand as replacements; the old nib was simply pulled out of the handle and a new one inserted. The tedious chore of constantly sharpening a quill became a thing of the past. In 1880, the English inventor James Perry improved the pen nib by cutting a hole at the top of a central slit and adding other slits on the sides to form a more flexible point.

Having to constantly dip a pen in ink to replenish the supply was tedious. Anyone old enough to remember learning to write in school with a dip pen and ink using the Palmer Method of push, pulls, and circles can attest to the fact that it was an ordeal. In 1884, an American inventor, L. E. Waterman, produced the first practical fountain pen. The fountain pen has a built-in rubber bladder to hold ink and is fed to the pen point by capillary action. In the ensuing years many different companies have produced fountain pens. Some are made to fit in a stand on a desk; others have a screw on cap and are carried in a pocket or purse.

Ballpoint pens first appeared in the late nineteenth century; however, because they did not function well, they were not popular with the public. The writing point of a ballpoint pen consists of a metal ball in a socket that freely rotates and rolls the ink from a reservoir onto paper. The ball in the early pens did not roll smoothly; they would skip leaving blank spots or leak ink leaving messy fingers. In the 1930s, a Hungarian living in Argentina, Lazlo Biro, patented a ballpoint pen, referred to as the "Biro," that used an oil-based ink. The Biro performed smoothly, without the skips that had plagued other ballpoints. By the 1940s, the much-improved ballpoint pen was widely accepted by the public. Although the fountain pen is still popular today, the ballpoint pen has become the most widely used writing instrument in the world.

Inkwells & Inkstands

As we enter the twenty-first century, we are deep in the age of the computer. Our method of communicating with the written word is now via E-mail. A typed message can be sent around the world as fast as it takes to hit the "send" button, and we can receive a reply just as fast. Communications have progressed beyond anyone's wildest expectations. In a few scant years we have moved from centuries of writing with a pen dipped in ink to the computer keyboard. The inkwell is now as obsolete as the horse and buggy. No longer are inkwells to be found in every home, hotel, shop, or other places of commerce. No longer does a traveler need to carry a small inkwell in his pocket. No longer does a mischievous schoolboy sitting in a desk behind a little girl dip her pigtails into the inkwell. The era of the pony express with its horse and rider pounding over rough turf through rain and snow to deliver the mail has faded from memory. A long and fascinating period of writing history has come to an end. Although none of us would choose to return to those olden days, we can take a look back through history and see how far we have come.

Ink, a fluid made of various substances, has been used with an instrument for writing for thousands of years. Therefore it is reasonable to expect that some sort of receptacle be created to hold this liquid. One of the earliest ink containers was probably a cow or oxen horn; the hollow interior would provide a natural reservoir. The "Inkhorn" was followed over the centuries by thousands of other vessels made in every size and shape, constructed from every kind of material imaginable.

Inkwells are individual containers that hold ink, whereas an inkstand holds one or more inkwells and possibly other accessories. In some old manufacturer's catalogs the term "inkwell" and "inkstand" was used interchangeably. In the eighteenth and nineteenth centuries, in England, an inkstand was referred to as a "Standish." Over time many accouterments were added to the inkstand; one or more of the following was incorporated into the design: a pounce pot or sander, a compartment for a seal, a box for extra nibs, a brush to clean ink from clogged pens, a candle holder, a stamp box, a match container, and a call bell to ring for a servant to post the letter. There are one or more holes in the top of inkstands to hold quills in an upright position to protect their sharpened points. When steel nibs came into use, pens could be stored horizontally, so pen racks were added to the stand.

Containers to hold ink were made in a vast array of shapes and styles, from the pastoral to whimsical. Animals of all sorts were greatly favored; everything from the aardvark to the zebra was depicted, and many of them had glass or ivory eyes. Children, cherubs, and imps were another popular motif. Items such as boots, buckets, hats, musical instruments, ships, shoes, and tree trunks were just a few items that were used to hold ink. Important figures from history were depicted. Napoleon was often favored. Inkstands were made to commemorate important dates and events. Souvenir inkwells were sold in great abundance, because they were small, easy to pack, and made attractive and useful gifts.

Inkstands have been crafted in brass, bronze, cut crystal, glass, gold, gutta percha, horn, leather, mother-of-pearl, paper maché, pewter, porcelain, pottery, sterling silver, silver plate, and wood. They have been gilded, painted, decorated with flowers, engraved, inlaid with ivory and mother-of-pearl, and some set with precious gems. Cabinetmakers made inkstands from pine, birch, mahogany, rosewood, and walnut burl to go with the furniture styles of the day. Silversmiths made inkwells in addition to flatware, tankards, bowls, candlesticks, et cetera.

Before the seventeenth century very few people could read or write. When a written document was needed, a professional scribe or scrivener, an educated man, was hired for the task. The Industrial Revolution started in England in the eighteenth century and eventually spread to other countries, bringing vast changes. People moved from a rural environment into cities to work in factories. For the first time children could attend free public schools on a regular basis, and the literacy rate increased significantly.

As more and more people acquired the ability to read and write, the need for inkwells increased greatly. In England two additional things occurred that increased the demand for inkwells: the invention of the steel pen nib, which eliminated the difficulties associated with feather quills, and Rowland Hill's invention of the adhesive stamp in 1837. Britain released the first postage stamp on May 1, 1840. It was referred to as the penny-black and featured the portrait of Queen Victoria. These proved to be very successful, and, by 1860, most countries had adopted the use of postage stamps. The United States issued its first official stamps in 1847: a ten-cent stamp with the portrait of George Washington, the first President of the United States, and a five-cent Benjamin Franklin stamp.

The demise of the inkwell began when Lewis E. Waterman, an American inventor, marketed the first practical fountain pen in 1884. By 1920, the fountain pen had become the pen of choice for most people. This was followed in the 1930s with the ballpoint pen. By the 1950s, the United States Post Office, one of the last holdouts, had replaced their outmoded dip pens and inkwells with ballpoint pens.

Containers for ink have been made in every corner of the world over many centuries. Some are quite rare and are now in museums, or in the collections of the very wealthy. However, thousands of inkwells are still available for the today's collector. They can be found at antiques shows and shops, auctions, flea markets, consignment shops, and garage sales. The prices range from just a few dollars to the very expensive. Rare ones may cost thousands of dollars.

Inkwells are a delight to collect; they are relatively small, so they take up little space, and the variety is so vast it is impossible to become bored with the subject.

To gaze upon an old inkwell and imagine the people who originally owned it is a pleasurable pursuit. Did this plain pewter inkwell once sit on a desk in the House of Burgesses in Williamsburg, Virginia? Did this beautiful French porcelain one with the delicate hand-painted flowers once belong to Madam Pompadour? A forty-niner may have used that sturdy counting house inkwell during the Gold Rush. A tiny traveler's inkwell with its tightly fastened lid may have been carried in the pocket of a Civil War soldier or possibly used on a sailing ship.

Travelers or Pocket Inkwells

With a ballpoint pen tucked in our pocket we are far removed from those olden days when writing was done with a quill dipped in ink. Travel in past years presented a special problem: how to carry ink and an inkwell on the back of a horse, a rolling ship, or a wagon train, without the mess of spilled ink? Small, two to three inch, traveling inkwells first appeared in the seventeenth century, as part of a desk or carriage box, that contained an inkwell, writing paper, and quills. By the eighteenth century, the small inkwell, with a securely locking lid, was carried in one's pocket. These small pocket wells were very popular and used extensively throughout the nineteenth century.

England, France, and Germany made vast quantities of pocket inkwells in a diversity of designs, using many different materials. Travelers were made from brass, bronze, glass, gutta percha, horn, pewter, rubber, silverplate, sterling silver, tin, wood, and sometimes from a combination of these materials. A popular form was brass or nickel silver covered with red, brown, or black leather, with a spring-loaded push button opener hidden underneath the leather. The word INK was often stamped on the lid in gold letters. These usually have two securely hinged lids: one on the outside and the other over the inkwell inside. Some have added accouterments such as a pen wiper brush, a tiny sander, a candle, or sealing wax holder.

In the United States, the Silliman Company, founded in 1832 in Chester, Connecticut, produced many writing accessories: pounce pots or sanders, seals, wafer containers, inkwells for school desks, and traveler's inkwells. The small "pocket inkstand" was put in production in 1854. Most of these were made in the shape of a cylinder or barrel, and constructed from boxwood, cocoa wood, or rosewood. A Silliman ad informs us: the glass ink bottle inside is "surrounded with an air chamber, so that the ink is not liable to freeze in cold weather, if the wells are kept corked." These travelers could be purchased with a screw on top or a bayonet catch. During the American Civil War (1861-1864), the Silliman pocket inkwell was part of a soldier's kit. Letters were written to love ones at home using a quill dipped in a pocket inkwell, relating the horrors of war fought in places like Gettysburg, Antietam, Bull Run, Shiloh, and other bloody battlefields.

Matching sets were made for two pockets: an inkwell and a duplicate container to hold matches to melt the sealing wax. The match holder has a hinged outer lid that snaps shut to close and an inner lid over the match compartment that has a ribbed surface on which to strike the match. A set is rather hard to come by, as most of the sets have been separated over the years.

Figural inkwells are highly prized by the collector. They were made in many shapes—a head, a violin case, a book, medical bags, a hat, etc. Many novelties were made as souvenirs; these may be from the Columbian Exposition (held in Chicago, Illinois, in 1893 to commemorate the 400th anniversary of Columbus's discovery of America), Niagara Falls, Paris, London, Rome, or any place a tourist might visit. These small inkwells were easy to carry home and made excellent gifts.

A rather unique item that originated in the Orient is a brass or bronze writing case that was carried tucked in a wide sash around the waist. It has a long tube for holding writing brushes or quills with an attached inkwell on one end.

Some collectors are interested exclusively in traveler's inkwells, while others consider them an interesting addition to their collection of standard inkwells and inkstands. These small, ingenious, interesting, often amusing containers for ink bring a different aspect and pleasure to the hobby of collecting writing memorabilia.

U. S. Design Patents & Their Time Frames

Year	Number	Year	Number	Year	Number	Year	Number	Year	Number
1843	1-14	1874	7,083-7,968	1905	37,280-37,765	1936	98,045-102,600	1967	206,567-209,731
1844	15-26	1875	7,969-8,883	1906	37,766-38,390	1937	102,601-107,737	1968	209,732-213,083
1845	27-43	1876	8,884-9,685	1907	38,391-38,979	1938	107,738-112,764	1969	213,084-216,418
1846	44-102	1877	9,686-10,384	1908	38,980-39,736	1939	112,765-118,357	1970	216,419-219,636
1847	103-162	1878	10,385-10,974	1909	39,737-40,423	1940	118,358-124,502	1971	219,637-222,792
1848	163-208	1879	10,975-11,566	1910	40,424-41,062	1941	124,503-130,988	1972	222,793-225,694
1849	209-257	1880	11,567-12,081	1911	41,063-42,072	1942	130,989-134,716	1973	225,695-229,728
1850	258-340	1881	12,082-12,646	1912	42,073-43,414	1943	134,717-136,945	1974	229,729-234,032
1851	341-430	1882	12,647-13,507	1913	43,415-45,097	1944	136,946-139,861	1975	234,033-238,314
1852	431-539	1883	13,508-14,527	1914	45,098-46,812	1945	139,862-143,385	1976	238,315-242,880
1853	540-625	1884	14,528-15,677	1915	46,813-48,357	1946	143,386-146,164	1977	242,881-246,810
1854	626-682	1885	15,678-16,450	1916	48,358-50,116	1947	146,165-148,266	1978	246,811-250,675
1855	683-752	1886	16,451-17,045	1917	50,117-51,628	1948	148,267-152,234	1979	250,676-253,795
1856	753-859	1887	17,046-17,994	1918	51,629-52,835	1949	152,235-156,685	1980	253,796-257,745
1857	860-972	1888	17,995-18,829	1919	51,836-54,358	1950	156,686-161,403	1981	257,746-262,494
1858	973-1,074	1889	18,830-19,552	1920	54,359-56,843	1951	161,404-165,567	1982	262,495-267,439
1859	1,075-1,182	1890	19,553-20,438	1921	56,844-60,120	1952	165,568-168,526	1983	267,440-272,008
1860	1,183-1,365	1891	20,439-21,274	1922	60,121-61,747	1953	168,527-171,240	1984	272,009-276,948
1861	1,366-1,507	1892	21,275-22,091	1923	61,748-63,674	1954	171,241-173,776	1985	276,949-282,019
1862	1,508-1,702	1893	22,092-22,993	1924	63,675-66,345	1955	173,777-176,489	1986	282,020-287,539
1863	1,703-1,878	1894	22,994-23,921	1925	66,346-69,169	1956	176,490-179,466	1987	287,540-293,499
1864	1,879-2,017	1895	23,922-25,036	1926	69,170-71,771	1957	179,467-181,828	1988	293,500-299,179
1865	2,018-2,238	1896	25,037-26,481	1927	71,772-74,158	1958	181,829-184,203	1989	299,180-305,274
1866	2,239-2,532	1897	26,482-28,112	1928	74,159-77,346	1959	184,204-186,972	1990	305,275-313,300
1867	2,533-2,587	1898	28,113-29,915	1929	77,347-80,253	1960	186,973-189,515	1991	313,301-322,877
1868	2,858-3,303	1899	29,916-32,054	1930	80,254-82,965	1961	189,516-192,003	1992	322,878-330,169
1869	3,304-3,809	1900	32,055-33,812	1931	82,966-85,902	1962	192,004-194,303	1993	332,170-342,817
1870	3,810-4,546	1901	33,813-35,546	1932	85,903-88,846	1963	194,304-197,268	1994	342,818-353,931
1871	4,547-5,431	1902	35,547-36,186	1933	88,847-91,257	1964	197,269-199,954	1995	353,932-365,670
1872	5,432-6,335	1903	36,187-36,722	1934	91,258-94,178	1965	199,955-203,378	1996	365,671-377,106
1873	6,336-7,082	1904	36,723-37,279	1935	94,179-98,044	1966	203,379-206,566	1997	377,107-388,584

U. S. Invention Patents & Their Time Frames

Year	Number	Year	Number	Year	Number	Year	Number	Year	Number
1836	1-109	1869	85,503-98,459	1902	690,385-717,520	1934	1,941,449-1,985,877	1966	3,266,729-3,295,142
1837	110-545	1870	98,460-110,616	1903	717,521-748,566	1935	1,985,878-2,026,515	1967	3,295,143-3,360,799
1838	546-1,060	1871	110,617-122,303	1904	748,567-778,833	1936	2,026,516-2,066,308	1968	3,360,800-3,419,906
1839	1,061-1,464	1872	122,304-134,503	1905	778,834-808,617	1937	2,066,309-2,104,003	1969	3,419,907-3,487,469
1840	1,465-1,922	1873	134,504-146,119	1906	808,618-839,798	1938	2,104,004-2,142,079	1970	3,487,470-3,551,908
1841	1,923-2,412	1874	146,120-158,349	1907	839,799-875,678	1939	2,142,080-2,185,169	1971	3,551,909-3,631,538
1842	2,413-2,900	1875	158,350-171,640	1908	875,679-908,435	1940	2,185,170-2,227,417	1972	3,631,539-3,707,728
1843	2,901-3,394	1876	171,641-185,812	1909	908,436-945,009	1941	2,227,418-2,268,539	1973	3,707,729-3,781,913
1844	3,395-3,872	1877	185,813-198,732	1910	945,010-980,177	1942	2,268,540-2,307,006	1974	3,781,914-3,858,240
1845	3,873-4,347	1878	198,733-211,077	1911	980,178-1,013,094	1943	2,307,007-2,338,080	1975	3,858,241-3,930,270
1846	4,348-4,913	1879	211,078-233,210	1912	1,013,095-1,049,325	1944	2,338,081-2,366,153	1976	3,930,271-4,000,519
1847	4,914-5,408	1880	233,211-236,136	1913	1,049,326-1,083,266	1945	2,366,154-2,391,855	1977	4,000,520-4,065,811
1848	5,409-5,992	1881	236,137-251,684	1914	1,083,267-1,123,211	1946	2,391,856-2,413,674	1978	4,065,812-4,131,951
1849	5,993-6,980	1882	251,685-269,819	1915	1,123,212-1,166,418	1947	2,413,675-2,433,823	1979	4,131,952-4,180,866
1850	6,981-7,864	1883	269,820-291,015	1916	1,166,419-1,210,388	1948	2,433,824-2,457,796	1980	4,180,867-4,242,756
1851	7,865-8,621	1884	291,016-310,162	1917	1,210,389-1,251,457	1949	2,457,797-2,492,943	1981	4,242,757-4,308,621
1852	8,622-9,511	1885	310,163-333,493	1918	1,251,458-1,290,026	1950	2,492,944-2,536,015	1982	4,308,622-4,366,578
1853	9,512-10,357	1886	333,494-355,290	1919	1,290,027-1,326,989	1951	2,536,016-2,580,378	1983	4,366,579-4,423,522
1854	10,358-12,116	1887	355,291-375,719	1920	1,326,890-1,364,062	1952	2,580,379-2,624,045	1984	4,423,523-4,490,884
1855	12,117-14,008	1888	375,720-395,304	1921	1,364,063-1,401,947	1953	2,624,046-2,664,561	1985	4,490,885-4,562,595
1856	14,009-16,323	1889	395,305-418,664	1922	1,401,948-1,440,361	1954	2,664,562-2,698,433	1986	4,562,596-4,633,525
1857	16,324-19,009	1890	418,665-443,986	1923	1,440,362-1,478,995	1955	2,698,434-2,728,912	1987	4,633,526-4,716,593
1858	19,010-22,476	1891	443,987-466,314	1924	1,478,996-1,521,589	1956	2,728,913-2,775,761	1988	4,716,594-4,794,651
1859	22,477-28,841	1892	466,315-488,975	1925	1,521,590-1,568,039	1957	2,775,762-2,818,566	1989	4,794,652-4,890,334
1860	28,842-31,004	1893	488,976-511,743	1926	1,568,040-1,612,699	1958	2,818,567-2,866,972	1990	4,890,335-4,980,926
1861	31,005-34,044	1894	511,744-531,618	1927	1,612,700-1,654,520	1959	2,866,973-2,919,442	1991	4,980,927-5,077,835
1862	34,045-37,265	1895	531,619-552,501	1928	1,654,521-1,696,896	1960	2,919,443-2,966,680	1992	5,077,336-5,175,885
1863	37,266-41,046	1896	552,502-574,368	1929	1,696,897-1,742,180	1961	2,966,681-3,013,102	1993	5,175,886-5,274,845
1864	41,047-45,684	1897	574,369-596,466	1930	1,742,181-1,787,423	1962	3,013,103-3,070,800	1994	5,274,846-5,377,358
1865	45,685-51,783	1898	596,467-616,870	1931	1,787,424-1,839,189	1963	3,070,801-3,116,486	1995	5,377,359-5,479,657
1866	51,784-6,0657	1899	616,871-640,166	1932	1,839,190-1,892,662	1964	3,116,487-3,163,864	1996	5,479,658-5,590,419
1867	60,658-72,958	1900	640,167-664,826	1933	1,892,663-1,941,448	1965	3,163,865-3,226,728	1997	5,590,420-5,704,061
1868	72,959-85,502	1901	664,827-690,384						

A Gallery of Inkwells

An owl with glass eyes is standing in front of a tree branch. He has silver and copper colored feathers. The head is hinged at the top of the wings. An applied disc on the bottom has a lion in a circle holding a vase. Inscribed "Meriden Silverplate Company. Quadruple plate #2518." Meriden, Connecticut. America, c. 1885. 5 1/2" tall. $400-600.

A wise old owl with painted eyes is standing on an open book. The inkwell is hinged in the back of the owl's head. America, c. 1900. Painted white metal. 2 1/2" x 2 7/8", 4" tall. $225-300.

An owl is standing on the end of a long feather that is a pen tray. The owl's head is hinged in the back. There is a mark with an S in a circle. G.E. Schutzt. Austria, c. 1890. Bronze. 3" x 15 1/4", 5 3/4" tall. $800-1,000.

An owl's head with glass eyes. A hinge is below the head in the back. White metal. c. 1900. 2 5/8" in diameter, 3 1/8" tall. $150-225.

The full figure of an owl is standing on a base with cut corners. The head is hinged in the back. A very pretty effect is created with a sheet of red isinglass under an ornate openwork grill on top of the base. France, c. 1900. Gilt metal. 3" x 3", 3 1/4" tall. $400-600.

An owl with glass eyes is hinged in the back of the head. Inscribed with an English registry mark that tells us this item was made February 16, 1850. 1X is on the bottom. Brass. 3" x 3 1/2", 5 1/2" tall. $400-600.

A standing owl has a hinge in the back of his head. A pen rack in the rear holds three pens with space at the bottom for a letter opener. The round brush is for cleaning clogged pen nibs. Europe, c. 1900. Brass. 3 3/4" x 4 5/8". $250-350.

An owl with wings spread to form a pen tray. The head is hinged. Marked on the bottom "Bradley & Hubbard #4536." Meriden, Connecticut. America, c. 1900. Nickel-plated brass head on an iron base with a brass finish. 6" x 9", 4" tall. $400-600.

An owl with glass eyes is standing on a round base. The head is hinged in the back. England, c. 1900. White metal. 2 3/4" in diameter, 4 7/8" tall. $200-250.

A large owl with glass eyes and nicely detailed feathers. The lid is hinged behind the owl's head. The spread wings form a tray for pens. Silverplate. 4 1/2" x 10 1/2", 2 1/4" tall. $400-600.

An owl is standing with one foot on a tree branch. The head is hinged in the back. On the lower right hand side is a small hollow tree stump that holds a quill or pen. England, c. 1880. Brass. 2 3/4" x 4", 4 3/4" tall. $275-325.

Black Forest hand-carved double inkstand. A pair of owls with glass eyes are hinged in the back of the head. The base is painted black and has two channels for pens across the front. Germany, c. 1910. 4 1/4" x 8 1/2", 5 1/2" tall. $250-450.

Black Forest hand-carved wooden owl with glass eyes. The lid is hinged in the back of the head. Germany, c. 1910. $150-250.

An owl with glass eyes is standing next to a tree stump that is the inkwell. The lid is hinged. The channel across the front is a pen rest. Hand carved in Germany, c. 1920. 3 1/2" x 4 1/2". $150-300.

A hand carved wooden owl with glass eyes is standing on a black base. The hinge is located in the back, halfway down the body. Germany, c. 1900. 3" x 4", 7" tall. $200-250.

A hand carved owl with glass eyes is standing on an open book. Hollow log inkwell. Germany, c. 1900. Walnut. 4 1/4" x 5 1/4", 4 3/4" tall. $200-250.

A strutting rooster is standing on an irregular shaped base. There is a hinge in the back of the rooster's head and the top half opens. The tree branch in the rear forms a pen rack. France, c. 1870. Bronze. 4" x 6", 7 1/2". $600-900.

Inkwell in the form of a rooster's head with a red comb and glass eyes. The head is hinged in the back. America, c. 1920. Painted white metal. 2 1/8" x 2 3/4", 3 1/2" tall. $175-275.

A rooster stands on each side of a central inkwell. The oblong base has a beaded edge and stands on four ball feet. America, c. 1900. 5" x 7 3/4", 4" tall. $325-425.

Inkstand with a conquering rooster crowing over a downed eagle. The two inkwells have hinged lids with finials. A tray for pens is across the front. The composition stands on six scroll feet. France, c. 1900. Signed "A. Bossé." White metal with a bronze patina. Base: 7 1/4" x 10 1/2". Height to top of rooster: 8". $750-950.

Tiny inkwell that was probably made for a child. On top of the birdhouse is a dove feeding her baby. There are birds above the feeding trays on the four sides. The top half of the roof is hinged. c. 1920. White metal. 1 1/8" x 1 1/8", 2" tall. $195-250.

A child's colorful inkstand with a baby duck on top of the loose cover. The ledges across the front hold three pens. Germany, c. 1925. White porcelain with yellow and red trim. 4 1/2" x 4 1/2", 3 3/4" tall. $150-250.

French bronze in the form of a swan house with a swan in the doorway. The roof is hinged. The house stands on four legs on a black base. c. 1880. 3 5/8" x 3 5/8", 4 1/2" tall. $300-400.

Doré bronze figure of a stork standing on a saucer base. The stork is hinged above the wings. France, c. 1865. 4" x 6", 6 1/2" tall. $600-900.

Exceptionally beautiful Louis fifteenth style chinoiserie inkstand. A bird is perched on a stump between tree branches that are embellished with colorful flowers. The inkwell has a hinged lid. A pen rest is across the front. France, c. 1850. The bird, stump, flowers, and inkwell are porcelain; the base, trees, lid, and pen rack are gilt metal. 4" x 7 1/2", 6 1/2" tall. $2,500-3,500.

Large beautiful bronze inkstand with a pair of colorful parrots perched on tree branches. Austria, c. 1910. 7 1/2" x 18", 8 1/4" tall. The three-piece set includes a rocker blotter and a letter opener. $6,000-8,000 set.

The rocker blotter and the letter opener go with the parrot inkstand. The blotter is rather large: 3 3/4" x 7 1/4", 4" tall. The letter opener is 12" long (part of set seen at bottom of page 22).

Lusterware in the shape of a colorful parrot with a hinged head. The tray in front holds pen nibs or other accessories. Marked "Made In Japan." c. 1930. 3 1/8" x 3 1/2", 5 1/8" tall. $195-295.

Inkstand with a parrot perched on a tree stump. The inkwell on the left side has a loose cover. In the back is a hole for a pen. America, c. 1920. Painted aluminum. 3 1/2" x 6", 4 3/8" tall. $200-300.

Cut crystal with the figure of a parakeet perched on the top. The brass lid is hinged. 3 1/4" in diameter, 5 1/2" tall. $500-800.

A quail hen with her two chicks, one on her back and the other one nestled by her side. The lid is hinged below the chick on her back. c. 1900. Vienna bronze. 2 3/4" x 4 1/4", 3" tall. $800-1,200.

A kingfisher is standing next to a ball shaped inkwell that rests on three ball feet. The lid is hinged. Painted white metal. 3" x 3 1/4", 3 3/4" tall. $150-200.

The parrot's head is hinged in the back. Painted white metal. 2 1/8" x 3". $400-500.

An eagle with spread wings is mounted on a pedestal and flanked by inkwells with hinged lids. The channel across the front holds pens or other writing accessories. c. 1890. Brass eagle and inkwells on a marble base. 6 3/4" x 14 1/4". $600-800.

A hand carved bird with glass eyes and an orange celluloid beak. The inkwell has a hinged lid. American folk art. c. 1930. 3" tall. $75-125.

A mother bird is perched on a hollow log and when the hinged top is opened a nest of eggs is exposed. American folk art. c. 1920. Hand carved wood. 2 3/4" x 5", 3 1/4" tall. $175-250.

An eagle with spread wings is perched on a branch above a leafy nest that holds an inkwell. America, c. 1900. Painted iron. 4 1/2" x 5 1/4", 5" tall. $275-400.

An eagle with glass eyes is perched beside a tree stump inkwell. Across the front is a ledge for a pen. Hand carved wood from the Black Forest of Germany, c. 1910. 4 1/2" x 5 1/2", 5 1/2" tall. $250-350.

A large eagle is standing with spread wings. His head is hinged. The forked twig is a pen rest. America, c. 1890. Silverplate. 4 1/2" x 7 1/4", 6" tall. $600-900.

A ferocious looking eagle with glass eyes and a black tipped beak is hinged in the back of the head. America, c. 1925. Painted white metal. 2 1/2" x 3", 3" tall. $200-300.

The head of an eagle with glass eyes is mounted on a marble base. The top of the head is hinged. France, c. 1900. Gilt metal head. 3" x 6", 3" tall. $350-450.

An eagle is standing on a cannon beside a large red cannon ball inkwell. Parian ware. France, c. 1890. 2 7/8" x 3 1/8", 5 3/4" tall. $700-900.

A white dove with a fanned tail has a blue ribbon around his neck. The head is hinged in front, just above the bow. c. 1880. Painted white metal. 4" wide (tail), 4" tall. $250-350.

The white dove sitting on a nest lifts off and exposes a storage cup for nibs or wax wafers. An inkwell and sander are in front. The covers are removable. England, c. 1890. 4 1/4" x 5", 3 1/4" tall. $125-175.

A pair of geese are standing on top of the inkwell lids that slide to open. A peacock is standing in the rear and there is a pen rack in the shape of a stag's horn. A channel for pen nibs is between the two inkwells. England, c. 1900. Bronze with touches of red. 4 5/8" x 5", 3 1/2" tall. $275-350.

This inkstand is part of an eight-piece desk set. Not shown are four blotter corners, a rocker blotter, a stamp box, and a letter stand. The inkstand has a pair of pheasants on the top. The two inkwells have hinged lids. America, 1915. Bronze. 4" x 9", 3 1/4" tall. Inkstand: $300-400.

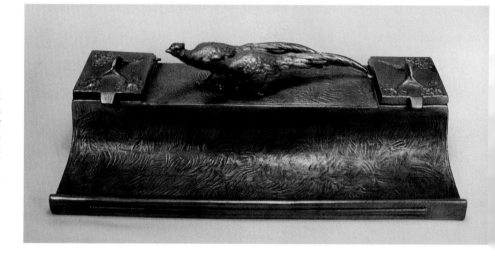

A kitten with glass eyes is sitting next to a large ball of yarn. The top half of the yarn is hinged. America, c. 1890. Silverplate. 3" x 4". $400-700.

A cat's head with glass eyes is sitting on a scalloped saucer base. The head is hinged. England, c. 1880. Polished bronze. 6 1/4" in diameter, 4 1/2" tall. $1,000-1,500.

White porcelain inkwell with a square cut lid. A pair of kittens in a basket are painted on the front. Embellished with blue and white pansies. England, c. 1900. The brass collar is hinged. 1 3/4" x 1 3/4", 2 1/2" tall. $150-250.

A carved wooden cat with glass eyes. The head is hinged. Germany, c. 1910. 3 1/8" tall. $250-350.

A cute kitten is attached to an odd shaped base. The swirl glass inkwell with a loose cover is fitted inside an open rail fence. A pen rest is on the front. America, c. 1900. Painted white metal. 2" x 3 3/4". $275-400.

A glass-eyed cat is peering over a pen tray. The head is hinged in the back. Germany, c. 1900. Carved walnut. 3 1/2" x 5 1/2", 3 1/8" tall. $225-295.

Whimsical cats with human bodies. The lady cat is wearing a yellow gown and the male cat is dressed like a soldier. A hole for a pen is on the front corners. Germany, c. 1920. Porcelain. 2 1/4" x 2 1/2", 4 3/4" tall. $450-550 pair.

An inkstand with a cat that has his eyes on a small mouse on the rim of the dish. The inkwell on the right side has a loose cover. The large irregular shaped tray holds pens or other writing accessories. Marked on the bottom "1605." French faience, c. 1890. 7" x 8 1/4", 2" tall. $800-1,000.

Pressed glass daisy and button pattern inkstand in the form of a chair. The loose cushion with a cat sitting on top is the inkwell cover. It is difficult to find a complete inkstand; the cat on the cushion cover is usually missing. The arms on the chair form a pen rack. America, c. 1900. 2 1/2" x 2 3/4", 4 1/2" tall. $200-300.

A dog is tied to a doghouse with a bowl by his side. The roof is hinged. Austria, c. 1880. Bronze. 4 1/8" in diameter, 3" tall. $300-400.

A cat is jumping through an opening in a curved rail fence. The two crystal inkwells have grooves on the sides that fit inside posts on the frame. The faceted lids are hinged. America, c. 1920. Brass. 4 1/4" x 6 1/4". $300-400.

A barefoot boy is feeding a dog that is chained to a doghouse. The roof is hinged. There is a candle holder for melting sealing wax, and a match container in back of the picket fence. Austria, c. 1880. Bronze. 3 1/2" x 7", 4 1/2" tall. $400-600.

This footed inkstand has a pierced gallery composed of berries and leaves that holds an inkwell and a stamp box. The swirl glass bottle has a hinged mushroom shaped lid. The two compartment stamp box has a dog standing on top of the hinged lid. There are hooks on the front to holds a pen. Marked on the bottom with a P in a diamond and "Pairpoint Mfg. Quadruple Plate 1997." America, c. 1885. 2 3/8" x 4 1/2", 3 1/8" tall. $350-550.

The dog's head is hinged in the back of the neck. c. 1900. Iron. 4 1/4" tall. $300-400.

A large dog with glass eyes has a small child leaning against him. The dog's head is hinged in the back. America, c. 1885. 3" x 6 1/2", 3 3/4" tall. Silverplate. $600-800.

A dog is sitting up in a begging pose. With his big glass eyes, laid back ears, and sweet expression, he is sure to win a titbit. The lid is hinged in back below the neck. England, c. 1880. Brass. 6 3/4" tall. $700-900.

Whimsical inkstand with a dog dressed in a suit and hat pushing a wheeled cart. The glass inkwell has a removable cover and sits in a square frame on top of the cart. Europe, c. 1880. Brass. 4" x 4 1/2", 3" tall. $300-450.

A puppy is on the top of the inkwell. The nursing bag beside him is connected to a tube in his mouth. The round inkwell has a hinged lid and the feet are shaped like dolphins. England, c. 1900. White metal. 2 1/4" in diameter, 3 1/2" tall. $200-250.

An extremely small pug's head with glass eyes and tiny ears. The head is hinged. Europe, c. 1880. White metal with a painted black base. 1 1/2" in diameter, 2" tall. $300-400.

A large dog with curly hair. The back of the head is hinged. America, c. 1900. Brass. 2 1/2" x 5", 2 1/4" tall. $150-195.

A dog with a curled tail is standing beside a boulder that is the inkwell. The lid is hinged. America, c. 1900. 2 3/8" tall. $195-225.

A brown and black dog with a hinge in the back of the head. Inscribed on the bottom "Knaff." Painted white metal. 2" in diameter, 3 1/2" tall. $200-300.

A seated dog with a long nose, long tail, and dropped ears. The dog's head is hinged in the back of the neck. c. 1885. Brass. 5 3/4" tall. $250-350.

A glass dog with a hinged brass collar. "Bonzo, Reg. No. 719074" is on the back, which indicates this inkwell was made in England in 1926. 1 3/4" x 2 1/4", 3 1/2" tall. $100-150.

Two dogs are sitting in front of a mailbox inkwell with a hinged lid. Diamond shaped base with #4610 on the bottom. The mail slot at the top of the letterbox is a pen rest. America, c. 1920. Cast iron. 5 1/2" x 5 1/2", 1 1/2" tall. $150-250.

A doghouse with a ferocious looking dog in the doorway. The roof is hinged. c. 1900. Silverplate. 2 1/8" x 2 1/8", 2 3/8" tall. $175-250.

A puppy with glass eyes is wearing a bib and has a pipe in his mouth. The head is hinged. "J. B. 683" is on the bottom. Silverplate. 5" x 5", 3 1/2" tall. $600-800.

Ball with a puppy on the top. Marked "J. B. 418." Jennings Brothers. America, c. 1920. Brass. 2" in diameter, 2 1/4" tall. $125-150.

A sad looking dog with dropped ears. The head is hinged in the back. The channel in front is for pens. Black Forest, Germany, c. 1900. Carved walnut. 5" x 7", 4" tall. $150-250.

A brown and white dog with dropped ears and large glass eyes. The dog's head is hinged in the back. Pen tray across the front. Marked "Buchin" on the bottom. Germany, c. 1910. Painted iron. 5 3/4" x 6", 4" tall. $300-500.

Baroque style inkstand fitted with a crystal inkwell. The matching faceted lid has a brass hinged collar. The figure of a dog is standing in front and a pen rack is in the back. Europe, c. 1900. 2 3/4" in diameter, 3 1/4" tall. $250-350.

The full figure of a dog sitting on the top of the inkstand. The crystal inkwell has a loose brass lid and is encased in a frame. A pen rest is across the front. The composition stands on four scroll feet. Probably France, c. 1900. Gilded brass. 3 1/4" x 5", 4" tall. The crystal inkwell is 1 3/8" square. $350-550.

Inkstand with two white dogs sitting in front of a hollow tree stump that is the inkwell. A small hole on the front of the base is to hold a pen. Marked "Staffordshire, England." $200-300.

Majolica inkstand with a white dog on a cobalt blue ottoman with yellow braids and tassel feet. The top half lifts off to expose two inkwells, a nib compartment in the middle, and channels for pens in the front and the back. Europe. 7 1/2" x 9 3/8", 4 1/2" tall. $1,500-1,800.

Bottom half of the majolica inkstand (on page 37) showing the compartments inside.

Footed inkstand with the figure of a dog in the middle. On the left side is a crystal sander and on the right the inkwell. Hinged lids with finials. Across the front is a channel for a pen. The platform skirt has beaded edges and is engraved with a leaf design. England, c. 1860. Silverplate. 7 3/4" x 11", 5" tall. $750-900.

A dog's head is in the center of a saucer shaped base. He is missing his glass eyes. England, c. 1900. Brass. 6 1/2" in diameter, 4 3/4" tall. $300-400.

A dog's head is mounted on a wooden base. The head is hinged. England, c. 1880. Brass. 6 3/4" in diameter, 4 1/2" tall. The dog is white metal. $300-400.

The figure of a running hound with a letter in his mouth is displayed on the top of the inkstand. A sander and two quill holes are exposed when the top is removed. France, c. 1880. Brass. 2" x 4 1/4", 3 1/2" tall. $600-700.

Running hound inkstand with the cover removed exposing the interior.

39

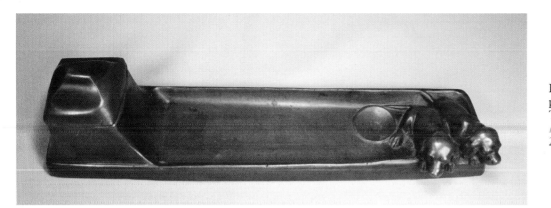

Pen tray with a pair of dachshund puppies and a bowl on the right side. The lid on the inkwell is hinged. America, c. 1910. Copper. 2 1/2" x 4 1/4". $200-300.

Hunting is the motif. A crystal inkwell with a faceted hinged lid is seated in a square frame in the back. The oval tray has openwork handles and a scene in the center depicting a dog baying over a downed deer. A dog's head is on the back rim. The scene is complete with crossed guns in the front. America, c. 1900. 7 1/4" x 9 1/2", 3 1/2" tall. $325-425.

The high-relief head of a spaniel is hinged in the back. The tray in front is a pen rest. The embossed heads of quails are displayed on the front and the sides, and a pair of quails are on the back. Embellished with acorns, oak leaves, and paw feet. America, c. 1885. White metal. 3 3/4" x 4", 3 1/2" tall. $350-500.

Beautiful rococo inkstand with open handles and a heavy border of scrolls and flowers. A pair of blown crystal inkwells are seated in frames on each side. In the center is a small box for pen nibs, wax wafers, or stamps. A spaniel is sitting on top of the removable lid. The two loose covers on the inkwells and the four feet are in the shape of flowers. England, c. 1850. Brass. 9 1/2" x 14 1/2". $800-1,000.

The dainty floral and ruffled design of this inkstand is in sharp contrast to the hunting scene depicted. Two hounds with snapping jaws in the bottom of the tray have cornered a boar. The two inkwells and pounce pot all have loose covers. Sitzendorf, Germany, c. 1830. Porcelain. 6 1/4" x 8 1/4", 6 1/2" tall. $4,000-5,000.

A bear with glass eyes is wearing a collar with a bow tie. The head is hinged. Hand carved in the Black Forest of Germany, c. 1910. 2 1/2" in diameter, 3" tall. $175-275.

Cute inkstand with the figures of a mother dog and five puppies emerging from a doghouse; five are on the front and one is on the left side. The roof is hinged. America, c. 1920. Brass. 3 5/8" x 5 3/8", 3 1/8" tall. $250-350.

A Black Forest carved wood inkstand in the shape of a brown bear with glass eyes. The bear is seated with an inkwell between his front paws. There is a white porcelain insert in the nickel-plated holder. Germany, c. 1900. 1 1/2" x 2 1/2", 3 1/2" tall. $175-350.

A pair of bear cubs are enjoying a stolen feast of berries and fruit. The bear on the right is throwing a tantrum because the bear on the left is downing the contents of the bowl. The tree stump inkwell has a removable cover. Russia, c. 1890. Porcelain. 4" x 5", 5 1/4" tall. $1,000-1,200.

Carved Black Forest bear with glass eyes. The head is hinged in the back. Germany, c. 1890. 2" x 3", 4 1/2" tall. $250-375.

Lusterware in the shape of a bear's head. The head is removable and there are two inserts inside. Channels on the top of the green base hold two pens. Marked "Made in Japan." c. 1920. $125-200.

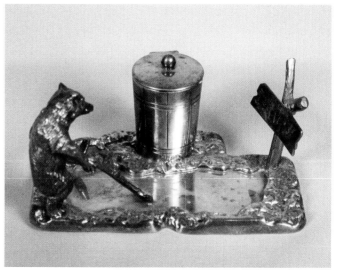

Whimsical inkstand with a bear on thin ice in front of a trash can. The sign on the tree stump reads "Danger." The lid on the can is hinged and opens to reveal an insert. The tray is edged with a decorative border. Inscribed "Pairpoint #5411." Pairpoint Mfg. Co. New Bedford, Mass. America, c. 1885. 3" x 5", 2" tall. $500-600.

Hand carved wooden inkstand in the shape of a kneeling camel. The saddle is hinged. Germany, c. 1880. 1 1/2" x 4 1/2", 2 3/4" tall. $225-295.

A large bear is standing on his hind legs clutching a hollow log in his front paws; inside the log is a pen. The bear's head is hinged in the back and opens to expose an insert. c. 1880. Brass. 9 5/8" tall. $700-1,000.

Hand carved figure of a camel on a round base with bark around the sides. The saddle is hinged. c. 1920. 3 3/4" x 4 1/2", 4" tall. $150-250.

Hand carved wooden inkstand in the shape of a kneeling camel. The saddle is hinged. 4 1/2" x 5", 3 1/2" tall. $125-150.

A lady is seated on the back of a kneeling camel. France, c. 1880. White metal with a bronze finish. 2 1/4" x 4", 3" tall. $200-300.

Camel shaped tray with a pressed glass inkwell seated at the top. Decorative loose cover. #4269 is on the base. America, c. 1920. Brightly painted cast iron. 5 1/2" x 6 1/2", 2 1/2" tall. $200-250.

Small kneeling camel with brightly painted furnishings. The fringed saddle is hinged. America, c. 1920. 2" x 5 1/8", 3 5/8" tall. $300-450.

Kneeling camel with a hinged saddle. America, c. 1920. White metal with a polychrome finish. 2 3/4" x 6", 4" tall. $300-450.

Kneeling camel with a decorative hinged saddle. America, c. 1920. Painted white metal. Small size: 3 3/4" x 7 1/2", 5 1/2" tall. $300-450.

A kneeling camel is wearing a brightly colored saddle decorated with fringes. The saddle is hinged. America, c. 1920. Painted white metal. Large size: 4" x 9", 9 3/4". $300-450.

A goat is pulling a wheeled cart that carries the inkwell. The white porcelain well is painted with red and blue flowers. The lid is hinged and attached to the cart by a chain. France, c. 1880. Brass on an alabaster base. 2 1/4" x 3 3/8". $250-350.

The top of the goat's head is hinged in the back. The goat is wearing a collar with four bells. France, c. 1890. Bronze. 3" in diameter, 4 1/2" tall. $400-600.

Two goats are in front and a goat herder is in the back. A square-cut crystal inkwell is fitted in a pierced frame. The mushroom shaped lid is hinged. The necks of the goats form a pen rest. The round base has a decorative border and stands on three ball feet. Inscribed on the bottom "James W. Tufts, Boston. Warranted Quadruple Plate 2855." America, c. 1880. 4 1/4" in diameter, 3 7/8" tall. $400-600.

The head of a ram is on the front and back of the inkstand. Supported on four legs with cloven hooves. Loose lid on a tan ceramic well. England, c. 1880. Brass. 4 7/8" x 4 7/8", 3 1/2" tall. $300-400

Charming inkstand with the figure of a doe resting with her fawn beside her. In the back is a pen rack fitted with a crystal inkwell. Embossed on the loose lid is the head of a girl with flowers in her hair. The composition stands on four claw feet. England, c. 1890. The inkwell is 1 7/8" x 1 7/8", 2" tall. The brass stand 6" x 8", 4 1/4" tall. $400-500.

A mule is pulling a wheeled cart with a clear crystal inkwell in the back. The faceted lid is hinged with a brass collar. Europe, c. 1900. Brass on an alabaster base. 2 1/8" x 3 1/4", 3 3/8" tall. $250-300.

The full figure of a donkey is standing in the back. The square crystal inkwell has a brass lid with the head of the woman embossed on top. On the left side is a pen rest. The base has four foliated feet. c. 1900. Brass. 4" x 5 3/4", 5 3/4" tall. $350-450.

A horse is pulling a two-wheeled cart with a crystal inkwell seated in the rear. The faceted lid has a brass hinged collar. Europe, c. 1900. Brass on an alabaster base. 2 3/8" x 8 3/4", 3 1/4" tall. $250-350.

A horse head with a halter. The top of the horse's mane and ears are hinged. England, c. 1890. Painted white metal. 3 1/2" x 5", 5" tall. $250-350.

Inkstand with the figures of two wolves standing on a rocky crag and a third one on the ground. They appear to be tracking a man; there are footprints across the top of the base that disappear behind the inkwell on the right. The square inkwell has a hinged lid. Austria, c. 1910. Bronze. 7" x 16", 8" tall. $1,500-2,200.

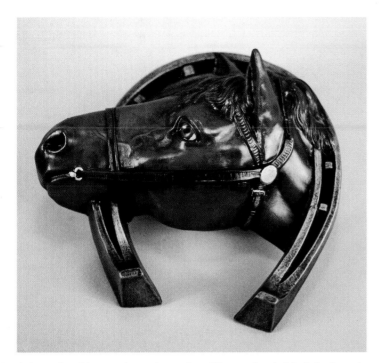

A horse head imposed on a horseshoe. The section behind the horse's nose and in front of the ears is hinged. Marked "Patent 35060," which is the patent number for 1901. America. Painted white metal. 4" x 4 3/4", 2 1/2" tall. $250-350.

A workhorse wearing blinders and a large collar. The head is hinged. Europe, c. 1870. Bronze. Base measures 2 1/2" x 3", 4 1/4" tall. $450-700.

Barometer inkstand on an octagonal base. The horseshoe nails form a pen rack. "Good Luck" is written across the top. America, c. 1875. Nickel-plated cast iron. 5 1/8" x 5 1/8", 4 3/4" tall. $200-300.

A horse's head is mounted on a black wooden base. The head is hinged. England, c. 1910. White metal. 6" x 7 1/4", 5" tall. $150-300.

A silverplated horseshoe with horseshoe nails forming a pen rack. The crystal inkwell has a hinged silverplated cover. The footed base is made of wood. England, c. 1900. 5" x 6 1/2", 3" tall. $300-400.

A pair of horses facing each other stand in the back. Seated in depressions on the top are two round glass inkwells with removable covers. Across the front is a divided water trough that is for a pen or other writing paraphernalia. America, c. 1880. White metal on a wooden base. 5" x 11 1/4", 6" tall. $500-700.

Art Deco inkstand with a cowboy riding a bucking bronco. The inkwell has a removable lid and there is a glass bottle inside. A pen fits in a holder in the back. Marked "Fount O Ink" on the stand as well as the bottle. Patent numbers under the lid and on the bottom of the bottle. America, 1912. Iron base. The rider and horse are white metal. $100-150.

A horse with his head lowered out of a stable appears to be interested in what is in the bucket below. The bucket inkwell has a hinged lid. America, c. 1910. White metal with a silver finish. $250-350.

Wooden horseshoe with a hinged riding cap in the center. England, c. 1900. Walnut. 3" x 4", 2 3/4" tall. $250-300.

Horseracing is the motif. A horse's head is imposed on a horseshoe and below is a jockey's cap that is the inkwell. The cap is hinged. England, c. 1910. Brass. 3 3/4" x 4 1/2". $150-200.

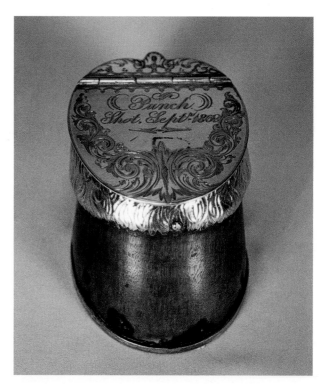

In England it was not unusual to memorialize a pet pony by having an inkwell made out of one of his hoofs. The top of this hoof has a silverplated hinged lid with the inscription "Punch shot Sept. 1869." 3 1/8" x 4 1/4", 2 1/2" tall. $400-500.

A horse is the motif of this inkstand. A horseshoe in the back has a horse's head at the top and four horseshoe nails to hold two pens. The base has four hoofs for feet. The two pressed glass inkwells have separate covers. On the bottom is inscribed "Pat. Dec. 11, 77. P. S. & W. Co." Peck, Stow, and Wilcox. Southington, Connecticut. America, c. 1885. Iron. 3 3/4" x 6 1/2", 5 1/4" tall. $300-400.

A horseshoe with the head of a horse at the top has four horseshoe nails to form a pen rack. The square base has four hoofed feet and is fitted with a glass inkwell with a loose lid. The horseshoe is a separate piece and has two prongs that hook it to the base. America, c. 1885. Cast iron. 3 7/8" x 3 7/8", 5 1/4" tall. $200-300.

Inkstand in the shape of a stag's head. Fitted with a pressed glass inkwell with a loose cover. The antlers form a pen rack. The base is composed of acorns and oak leaves. America, c. 1900. Painted aluminum. 4 1/2" x 6", 5 1/2" tall. $150-250.

A patterned glass inkwell is attached to the figure of a reclining deer. The brass cover is loose. The deer's horns form a pen holder. Marked JNA. Austria, c. 1890. 3" x 3 1/2", 3" tall. $400-500.

Stag and hounds. The two glass inkwells have funnel openings. The hooks on the front hold a pen. The oblong base stands on four feet. Bradley and Hubbard, West Meriden, Connecticut. America, c. 1900. Iron with a bronze finish. 5" x 9", 5" tall. $250-400.

Large impressive inkstand with the full figure of a majestic stag standing on a rough base. The inkwell on the right has a hinged lid. A pen channel is in front. Europe, c. 1880. Bronze. $1,200-1,400.

An antlered stag in high relief is on the front of this Art Nouveau inkwell. The hinged lid has an oak leaf and acorn handle. Additional acorns and leaves decorate the irregular shaped base. Inscribed on the bottom "Trade Mark J. B. Signifies The Finest 503." Jennings Brothers, Bridgeport, Connecticut. America, c. 1910. Silverplate. 3 1/2" x 4 1/2", 3 1/2" tall. $150-250.

A gravity fed glass beehive has a brass font in center front with a removable cover. Two brass deer heads with elongated antlers form a pen rack. c. 1870. 4" in diameter. $400-500.

A clear crystal inkwell is seated in a recess on top of the oak base. Loose cast iron cover. The antlers on the stag's head form a pen rack. England, c. 1910. 5 7/8" x 6 1/8", 4" tall. $250-350.

Silverplated inkstand with a stag's head in the rear; the antlers form a pen rack. A pair of round cut crystal inkwells are fitted in frames on each side. An engraved leaf design is on the hinged lids, the top of the stag's head, and around the edges of the tray. c. 1900. 8" x 9 3/4". $250-350.

The foxhunt is the motif. The head of a fox sits on top of the hinged lid. A stirrup with crossed riding crops forms a pen rest. A wreath of oak leaves and acorns encircles the base. Austria, c. 1880. Iron base with a painted white metal head. 5 1/2", 4" tall. $500-700.

A stag's head is mounted on top of the hinged lid. The base is encircled with a wreath of oak leaves and acorns. A stirrup is in the middle of the base; a pair of crossed riding crops form a pen rest. Austria, c. 1880. Iron base with a painted white metal stag. 6" x 6 1/2", 6" tall. $500-700.

A fox head with glass eyes is in the center of an upside down umbrella. The head is hinged in the back. Germany, c. 1880. Bronze patina. 5 1/2" in diameter, 3 1/4" tall. $450-700.

The head of a buffalo is mounted on a black base. The head is hinged in the back. America, c. 1890. Painted white metal. 1 3/4" x 3", 3" tall. $300-400.

American bison, commonly called a buffalo. The buffalo with a hump on his shoulder and a horned head, once grazed across the United States in great numbers, estimated as many as 60,000,000. Until the arrival of Europeans, the buffalo herds supplied the Plains Indians with meat and hide. With the ensuing westward migration of white civilization, the wanton slaughter of the animal decimated the herd to near extinction.

The full figure of a buffalo forms the inkstand. The buffalo's hump is hinged. #4253 is on the bottom. America, c. 1880. White metal with a bronze finish. 2 1/2" x 8", 5" tall. $600-900.

The figure of an American buffalo is standing on an irregular shaped tray. The inkstand is fitted with a crystal inkwell with a hinged lid. America, c. 1900. Brass. 4 1/4" x 6 1/2". $200-250.

The full figure of an American buffalo is standing beside a hollow tree stump inkwell. America, c. 1900. White metal. 2 1/2" x 5". $250-300.

Inkwell made out of a deer hoof. Nickel-plated hinged lid. America, c. 1910. 2" x 3 1/2", 5 1/2" tall. $125-175.

Hand carved wooden inkstand with the figure of an antelope standing on the top. On the right side is a glass inkwell with a separate wooden cover. The hollow log across the front holds pens. Germany, c. 1890. 3 1/2" x 6", 5 1/4" tall. $125-175.

A carved wood antelope with glass eyes. He has a white muzzle and a pair of black horns that curve back and down. The head is hinged. Germany, c. 1890. 3 1/2" wide, 5 1/4" tall. $300-400.

A carved wood antelope with glass eyes set in a black mask. His horns curve up and back. The head is hinged. Germany, c. 1890. 3 1/2" wide, 5" tall. $300-400.

The figure of a lion is attached to a tree stump inkwell. The top of the stump is hinged. Austria, c. 1900. Painted white metal. 3 1/2" x 3 1/2", 2 1/4" tall. $200-300.

Hand carved head of a chamois, a small antelope. He has short black horns and glass eyes. The head is hinged. Germany, c. 1880. 1 1/4" x 2", 3" tall. $300-400.

The head of a roaring lion on a black wooden base. Europe, c. 1910. White metal. 5" in diameter, 4 1/2" tall. $225-300.

Impressive inkstand with the regal figure of a lion standing on the top. The two crystal inkwells have hinged silverplated lids decorated with large flowers. The footed base has a face in the front and ornate scrolls around the sides. The oblong channel holds a pen. Probably English, c. 1900. Silverplated. 6 1/4" x 11 1/2", 7" tall. $900-1,200.

A bizarre inkstand. A roaring lion's upper body is rolled over and rests on his stomach. With crossed paws on top, back paws, and tail on the front, a tray is formed to hold pens. Europe, c. 1900. Painted iron. 9" x 10 1/2", 4 1/2" tall. $700-1,000.

A ferocious looking lion with ears laid back and teeth bared. His head is hinged in the back. Europe, c. 1900. Painted white metal. 6" x 6", 4 3/4" tall. $300-500.

An inkwell in the form of a roaring lion's head. The head is hinged in back. Europe, c. 1880. Painted white metal. 3 1/4" x 3 1/2" 3 3/4" tall. $225-300.

A pair of pouncing lions are facing outward on each side. The two inkwells have hinged dome-shaped lids with finials. A woman's face is on the backplate. The large oval tray in front holds pens and accessories. A bouquet of flowers is on the front and a leafy border encircles the base. Germany, c. 1890. Brass. 7 3/4" x 12 1/2", 3 1/2" tall. $400-600.

A circus elephant has a monkey riding on his head. The headscarf is hinged in the back. The monkey is the finial. America, c. 1870. White metal. 3" x 6 1/2", 4" tall. $250-350.

This is a delightful inkwell. An elephant with real ivory tusks is sitting on his rump with his trunk curled upward. The decorative cover on his head is hinged in the back. China, c. 1870. White metal. 4" x 4 1/2", 4 3/4" tall. $1,500-2,000 (rare).

Inkstand displaying the full figure of an elephant. The irregular shaped base has an inkwell in the form of a tree trunk on the right side. The lid is hinged. Marked "Austria" on the bottom. c. 1890. Bronze. 3 1/2" x 9 1/2", 3" tall. $400-600.

The figure of an elephant with his trunk raised showing his white tusks. The square inkwell on the left has a hinged lid. The long channel holds pens. America, c. 1920. The elephant is white metal; the base and inkwell are marble. 6 3/8" x 9 7/8", 6 1/4" tall. $300-400.

A circus elephant with a monkey riding on his head. The decorative headscarf, with the monkey finial, is hinged. This inkwell was advertised in a Meriden Britannia and Company Catalogue in 1877. America. Silverplate. The base measures 4 3/4" in diameter, 5 1/2" tall. $400-600.

The head of a long tusked walrus. The head is hinged in the back. c. 1900. Brass. 3 1/2" in diameter, 4" tall. $400-600.

An elephant is standing on a turtle while supporting a globe on his back. The globe is hinged. The composition stands on a round saucer shaped base. c. 1880. Brass. 5 1/2" in diameter, 6 1/4" tall. $400-600.

A pair of mice appear to be ready for a feast. A mouse is sitting on the top of a stack of crackers and cheese and another one is crouching on the side. The top cracker is hinged. Marked on the bottom "Deposé 108. 17 Adms & D." France, c. 1885. Vienna bronze. 2 12" x 3 7/8", 2 1/8" tall. $800-1,200.

This large inkstand depicts a pair of seals frolicking in ocean waves. The inkwell has a hinged lid and is located on the right side. Signed "Teariey." c. 1880. Bronze. 6 1/2" x 12", 3 3/4" tall. $800-1,000.

A rat has found himself a tasty meal, a succulent parsnip. The top of the parsnip is hinged. Austria, c. 1885. Bronze on a marble base. 3 1/4" x 4 5/8", 4" tall. $600-800.

A boar's head is sitting on a tree branch in a bed of oak leaves. The eyes are made of glass. The hinge is in the back of the head. England, c. 1900. Painted white metal. 4 1/4" x 7", 4 1/2" tall. $600-800.

A rabbit is sitting on the top and another one is on the ground below. Carrots are in a dish on the right side. Hinged cover. A pen holder is across the back. England, c. 1880. The slate base is 3 1/2" in diameter. 3 1/2" tall. $500-700.

This turtle was once alive. The shell has been cut to make a lid that is hinged in the back. Inside are two milk glass inkwells with removable brass lids. The turtle has glass eyes. France, c. 1900. 3" x 6 3/4". $750-950.

An alligator with an open mouth is attached to the side of a tree stump inkwell. America, c. 1920. Painted white metal. 3" x 3 1/2", 1 3/4" tall. $200-300.

A clear crystal inkwell is attached to a bronze lizard that has emerald green eyes. The faceted lid has a hinged bronze collar. England, c. 1850. 1 3/4" x 1 3/4", 4 1/2" tall. $450-650.

A brass lizard is curled around the front of a crystal inkwell that is attached to a brass base. The lid is hinged with a brass collar. England, c. 1900. 2 1/2" x 2 1/2", 3 3/8" tall. 5 1/2" (includes the lizard). $450-600.

A crab with large pincher claws. The shell is hinged in the back. England, c. 1880. Brass. 6 1/2" x 7", 2 1/2" tall. $700-1,000.

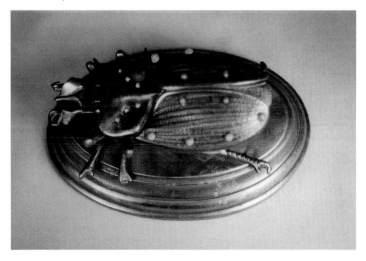

A beetle with turquoise eyes and coral and turquoise jewels set in his back. The wings are hinged in the front. Austria, c. 1880. Bronze. 3" x 4 1/2", 2" tall. $350-450.

A bronze crab. The shell is hinged and opens to expose a bronze insert. Japan, c. 1900. 5" x 8". $500-700.

Inscribed on top of the crab's shell "World's Fair. St. Louis 1904." America. Aluminum. 5 1/2" x 6", 1 7/8" tall. $150-250.

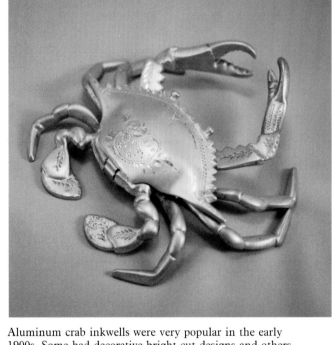

Aluminum crab inkwells were very popular in the early 1900s. Some had decorative bright cut designs and others were engraved to commemorate special events such as fairs or conventions. This crab has a bright cut floral and leaf design on the body and claws. The shell is hinged. America, c. 1905. 5 1/2" x 6 1/2", 1 3/4" tall. $150-250.

Brass crab with a hinge in the back of the shell. America, c. 1910. 4 5/8" x 6 1/2". $300-400.

Aluminum crab with a hinged shell. Single insert. Engraved "Brewmasters Convention, Baltimore 1907." America. 5 5/8" x 5 7/8". $150-250.

Art Nouveau inkstand displaying the lovely figure of a woman with fairy wings. The inkwell is encased in the tulip on the right side; the hinged lid is decorated with flowers. France, c. 1895. Bronze. The marble base measures 5" x 7". The height is 4 1/2". $400-600.

Seashells are the motif. The top shell has a snail finial and is hinged; inside are two glass inserts. Flowers and ivy leaves enhance the design. Europe, c. 1900. Bronze. 6" x 8", 2 1/2" tall. $350-500.

A mermaid, or siren, is a mythical woman with the head and body of a human and the tail of a fish. She was reputed to sing beautiful songs that lured mortal men to live with her in the sea. A beautiful bronze inkstand with a mermaid on one side and an octopus on the other; a reverse painting of the sea is in between. The head of the octopus is hinged. The lower front is in the shape of an oyster shell. Inscribed "Real Bronze. Geschutzt." Germany, c. 1900. 8" x 9", 2 1/2" tall. $1,200-2,000.

Art Nouveau inkstand with the figure of a woman curled up on a lily pad with her arms around a flower pod. The top of the flower is hinged. Number 4 is on the bottom. America, c. 1900. Silverplate. 3 3/4" x 6", 3 1/2" tall. $300-400.

Art Nouveau inkstand that displays a mermaid basking on a bed of rocks with the sea lapping around her. Flanked by a pair of cut crystal inkwells fitted with loose covers that have the figure of a fish for a finial. A pen channel is in front. Germany, c. 1890. Gilt white metal. 4" x 8". $400-600.

Art Nouveau inkwell with the head of a girl whose long flowing hair forms handles on the sides. The top of the head is hinged in the back. America, c. 1880. White metal with a bronze finish. 5 1/4" x 6 1/2", 3 1/4" tall. $125-175.

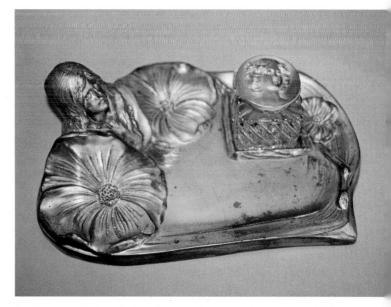

Art Nouveau inkstand that displays the head of a woman with long flowing hair nestled between two large flowers. A swirl pattern glass inkwell with a separate lid is seated in a frame on the right side. France, c. 1890. Bronze with a gilt finish. 4 3/4" x 6 1/2". $350-450.

Art Nouveau inkwell in the shape of a woman with long flowing hair. She is wearing a hinged flower garland. America, c. 1910. White metal with a gold finish. 2" x 2 3/4", 2 3/4" tall. $150-225.

The head of a woman wearing a large bonnet that is tied under the chin. The tray is in the shape of a four-leaf clover. Marked #278. America, c. 1900. Iron. 6" in diameter, 1 3/4" tall. $175-300.

A black haired girl in a red robe is sitting on a mat with a green inkwell between her crossed legs. There is a hole in the top of her head to hold a pen. Marked "HEU BACH 11738." Germany, c. 1920. 2 1/2" x 4", 4 5/8" tall. $450-550.

Egyptian motif. The head of a Pharaoh is on a footed bowl. The head is hinged. Replicas of coins are imbedded around the saucer shaped base. France, c. 1870. Bronze. 6" in diameter, 5" tall. $600-800.

Arabian scene with a woman seated on a camel while a man is standing in front of her. The inkwell is the round dome on the right. Signed on the front "Oran" and on the back "Ouveh." France, c. 1930. White metal. 3 1/2" x 7 7/8", 6 5/8" tall. $200-300.

Egyptian style inkstand with the bust of Cleopatra. The head is hinged. The saucer is decorated with a pair of lions face to face interspersed with circles. France, c. 1870. Bronze. 8" in diameter, 7 1/2" tall. $600-800.

Inkstand in the form of a woman paddling a boat. The inkwell is on the left side and has a hinged lid. Although the girl is attired in Dutch style clothes, including a bonnet and wooden shoes, this is believed to be of French origin. c. 1900. Gilt over brass. 5" x 9 1/2", 4 3/8" tall. $200-300.

Art Nouveau inkwell with the figure of a woman on the hinged lid who appears to be bathing in a lily pond. Water lilies decorate the lower half. America, c. 1900. White metal with a bronze patina. 2 1/4" in diameter, 3 1/4" tall. $200-300.

The statue of a tall slender woman dressed in a long gown and wearing beads is typical of the Art Deco period. She is flanked by pressed glass inkwells with loose covers. Embossed on the front "GLAZO." c. 1925. Gilded white metal. 2 3/8" x 6", 7 3/4" tall. $200-300.

A woman's head has ivory eyes. The flat hat is hinged in the back. Germany, c. 1850. Carved wood. 2 5/8"x 3 5/8", 5" tall. $375-475.

The figure of a Dutch girl in a blue dress, white apron, and wooden shoes stands beside a square crystal inkwell. The round lid is faceted and has a nickel-plated hinged collar. Holland, c. 1900. White metal. 1 1/2" x 3", 3 5/8" tall. $125-175.

Porcelain inkwell in the form of a woman wearing a large hat and a dress with dots. A feather is in the quill hole. The inkwell is under the woman's chin. Marked with an anchor in a circle "Bonlogne-Sur-Mer" and below "France." c. 1920. 2 5/8" x 4 7/8". $500-700.

Art Deco inkwell in the shape of an Indian woman sitting with crossed legs. She is wearing a turban with a jewel in the center. #6502 is on the bottom. Germany, c. 1920. 3" x 3 1/2", 5 1/2" tall. $250-350.

Mr. & Mrs. Carter Inx were advertised in the *Saturday Evening Post*, September 12, 1914. These cute ink bottles were given away with a 25¢ coupon. Made in Germany for Carter Ink until World War I, and later in the United States. Patented January 6, 1914. Brightly colored porcelain. Ma is 2 1/2" in diameter, 3 1/2" tall. Pa is 2 1/2" diameter, 3 3/4" tall. These were also made in different sizes. $250-400 pair.

Baker chocolate lady carrying a cup of hot chocolate on a tray. She is hinged in the back below her arms. America, c. 1900. Painted white metal. 4 1/2" tall. (This tiny inkwell is a double collectible, it is also desired by the advertisement collector.) $1,000-1,500 (rare).

A cut crystal inkwell is seated in a frame on the top of a wheelbarrow being pushed by young boy. The faceted lid is hinged with a brass collar. Europe, c. 1880. The boy is silverplated; the wheelbarrow is cast brass. 1 1/2" x 4 1/4", 2 7/8" tall. $300-450.

A little girl is wearing a bonnet with a bow tied under her chin. The hinge is on the side of her face. France, c. 1880. Bronze. 4 1/2" x 5 1/2", 2 1/2" tall. $450-550.

A child's face is framed in a bonnet. The face is hinged on the right side. Stands on three feet. France, c. 1890. Heavily decorated brass. 4 3/4" x 7". $250-350.

A cherub is seated on a pedestal in the rear. A square crystal inkwell fits in a frame on the top of the round wooden base. The faceted lid has a hinged collar. Two upright leaves form a pen rack. France, c. 1890. White metal. 7 1/4" in diameter, 7 1/4" tall. $600-800.

Inkwell with a winged cherub on the top. The lid is hinged. The drum shaped base is resting on four dolphin feet. England, c. 1900. White metal. 2" x 2", 3" tall. $200-250.

A graceful openwork background for a fairy and a cherub seated on the upper right side. The two inkwells are fitted with dome shaped hinged lids. Cattails and flowers enhance the decoration. The irregular shaped base rests on four scroll feet. France, c. 1885. Gilded brass. 5" x 10 1/2", 5 1/4" tall. $750-950.

A child holding a toy is resting on two pillows. The bed is hinged at the mattress line. Probably France, c. 1830. Bronze on a red marble base. 3 3/8" x 5 1/8", 4 1/2" tall. $500-700.

A perky little clown has a pot between his feet that holds a pen. The top half lifts off to open. France, c. 1900. Colorful porcelain. 3 3/4" x 4 5/8", 5" tall. $350-450.

A clown is on his back with his feet up in the air. The inkwell is in his chest and has a loose cover with a green ball finial. A hole in the sole of his shoe holds a pen. Marked on the bottom "Becquera Aladin France. Made in France." c. 1910. 3 1/2" x 6 1/4", 7 1/4" tall. $600-800.

Pierrot the white-faced clown is wearing a ruffled collar and a tight black cap. His open mouth has a pen wiper inside. The head forms a loose cover. A hole on the side holds a pen. Marked "Aladin Made In France." c. 1920. 3" in diameter, 3 3/4" tall. $300-500.

A young boy in a yellow and red clown costume. The top of the hat is open and holds pens. The inkwell is exposed when the hat is removed. Marked "Germany" on the bottom. c. 1920. Porcelain. 2 1/2" in diameter, 4 1/4" tall. $150-250.

A clown is wearing a costume covered with red flowers. The hole on the side holds a pen. Marked "A. L. Made in France. MADle ADNET." c. 1930. Porcelain. 2 3/4" x 4", 7" tall. $200-300.

A little schoolboy is sitting on an overturned bottle marked INK. He appears perplexed by the fact that his hand, leg, and clothes are smeared with ink. Japan, c. 1930. Ceramic. 1 1/2" x 2", 4 7/8" tall. $95-125.

Imp or demon with a devilish looking smile is standing on two paw-footed legs. The hinged lid is in the shape of a leaf with a frog finial. A hole in the left ear holds a pen. c. 1890. White metal with a bronze finish. 4 1/2" x 4 1/2", 4" tall. $195-295.

A devilish looking head sitting on two paw-footed legs. The top of the head is hinged. The ears have holes in them to hold pens. c. 1890. Painted white metal. 2 1/2" x 3", 2 1/2" tall. $195-295.

Imp or demon that stands on two paw feet. The head is hinged. The ear openings hold pens. c. 1890. Bronze. 3" x 3 1/2", 4" tall. $195-295.

A brownie sitting on a tree stump is surrounded by seven toadstools. The tree stump has a lid that is hinged in the back. In fairy tales, a brownie is an imp or goblin, who does housework and good deeds at night while people are sleeping. America, c. 1890. Bronze. 4" in diameter, 4 3/4" tall. $300-400.

A smiling Cornwall fairy is sitting astride a mushroom with a snail on top. The mushroom cap is hinged. England, c. 1920. Brass. 1 3/4" x 2 1/4", 2 1/4" tall. $150-250.

A man is sitting on a tub holding his head with his left hand and a cane in his right. A hinge is located in the back of his neck. England, c. 1900. Bronze. 2 3/8" in diameter, 4" tall. $300-400.

An English gentleman sitting on a stile is attired in nineteenth-century clothes complete with a top hat. Behind the railing is a single pen rack. The figure is hinged in the back and opens just below the top button on his waistcoat. England, c. 1880. 3 1/2" x 3 3/8", 4 3/4" tall. $700-800.

A man wearing a flat hat is smoking a long pipe. The hat is hinged in the back. Germany, c. 1900. White metal. $300-400.

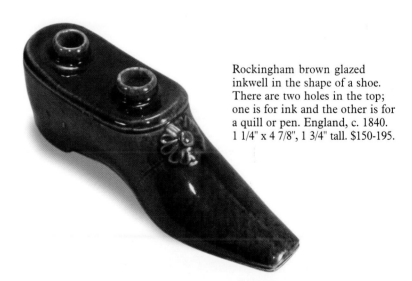

Rockingham brown glazed inkwell in the shape of a shoe. There are two holes in the top; one is for ink and the other is for a quill or pen. England, c. 1840. 1 1/4" x 4 7/8", 1 3/4" tall. $150-195.

A man with glass eyes is wearing Ben Franklin style glasses and a hat. The hat is hinged. France, c. 1885. Bronze. 3 1/4" in diameter, 4 1/8" tall. $500-700.

The figure of John Bull, a character depicted in English political satire as the personification of an Englishman, a portly, prosperous citizen. His head is hinged. England, c. 1860. Painted iron. 6" x 6 1/2", 4 3/4" tall. $400-500.

A man with a long goatee is wearing a pointed hat that is hinged in the back. A pen tray is across the front. Europe, c. 1900. Painted white metal. 5 3/4" x 7 3/4", 5 1/2" tall. $300-450.

A Roman soldier is wearing a laurel leaf hat and a high collar. The hinge is located in the back at the collar line. Marked #4891. Europe, c. 1910. Painted white metal. 4 1/2" x 5 1/4", 5 1/2" tall. $300-450.

The head of a man wearing a large feathered hat and a ruffled collar. A pair of swords are on the sides. The hat is hinged. Europe, c. 1890. White metal. 6 1/2" x 7", 2" tall. $300-500.

A bearded soldier in the French Foreign Legion. The cap is hinged. Inscribed on the lower front "UN POILU 1914 1916." Painted white metal. 2 3/4" x 5 3/8", 4 3/4" tall. $300-500.

A winged dragon is on the back with his tail coiled around the right side. The finial on the loose cover is the head of a woman. On the lower front is the face of a bat. On the bottom "Patented Mar. 5, 1901." America. 3 3/4" x 4 3/8", 4" tall. $300-400.

A clown wearing a smooth cap and a ruffled collar. The cap is hinged in the back. His open mouth held a pen wiper at one time. France, c. 1900. Brass. 4 3/8" wide, 3 1/2" tall. $300-400.

The head of a cavalier wearing a large hat. The feather is the hinged cover on the inkwell. Marked on the bottom "Deposé #118." France, c. 1900. Iron. 4 1/2" x 5 1/2", 1 1/2" tall. $250-400.

The figure of hooded monk reading a book. The head and shoulders are hinged. England, c. 1900. The base measures 2 1/2" in diameter, 4" tall. $250-350.

A man with a long pointed beard is wearing a Robin Hood style hat. The hat is the hinged lid on the inkwell. Marked on the bottom "J. B. 344." Jennings Brothers, Bridgeport, Connecticut. America, c. 1900. White metal. 2 3/8" x 4", 3 1/2" tall. $200-300.

A winged cherub is sitting on a pedestal flanked by crystal inkwells that have loose covers. A pen rest is in the front. France, c. 1880. Bronze. 4 7/8" x 5", 5 3/4" tall. $350-500.

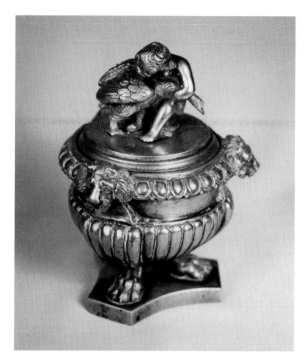

Urn on three paw feet. Three lion heads decorate the sides. A boy wrestling with a goose is on top of the removable cover. France, c. 1890. Bronze. $300-450.

A laughing black boy is wearing a floppy hat. The boy's head is hinged. America, c. 1900. Iron with a bronze finish. 3 1/4" x 6", 2 1/4" tall. $300-400.

The figure of a friar with a cup in his hand and keys hanging from his belt. The figure is hinged at the waist. Europe, c. 1900. White metal. 2" in diameter, 4 1/2" tall. $225-300.

A man in armor is peering through the grill on his plumed helmet. Armor was worn in battle to protect a man from the blows of a sword, spear, or arrows. This figure is referred to as "Kaiser Bill." Germany, c. 1910. White metal with a bronze patina. 4" x 4 1/4", 5 1/2" tall. $350-500.

View of the man in armor with the hinged lid open showing a grumpy looking warrior.

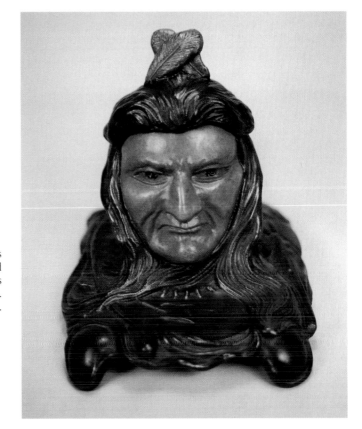

The head of an Indian with two feathers in his hair. On one side of the base is a tomahawk and on the other a bow and arrow. The head is hinged. America, c. 1900. Painted white metal. 3 3/4" x 4", 4" tall. $350-450.

An Indian in a large feather bonnet. He is wearing earrings and has bear claws around his neck. The headdress is hinged. There is a tray across the lower front for pen nibs. Embossed on the back is an emblem of three arches and a snake. America, c. 1900. Painted white metal. 6 1/2" x 7 1/2", 4 1/2" tall. $500-700.

The head of an Indian wearing a feather bonnet is a souvenir inkwell. On the front of the feathers is an emblem with a scenic picture of "Crater Lake" Oregon. The hinged feathers are the inkwell lid. America, c. 1910. Painted white metal. 3" x 3 1/2", 4" tall. $200-250.

The cap on the boy's head is hinged. America, c. 1880. White metal. (This inkstand was also made in bronze.) 2 1/2" x 5 3/8", 5 1/2" tall. $225-300.

A hunter with a gun in his right hand is standing beside a hollow tree stump inkwell. A smaller stump on the right side is a pen holder. Germany, c. 1880. Porcelain. 2" x 2 1/2", 4" tall. $275-325.

A French bronze inkstand with a barefoot boy wielding a scythe. On the right side is an inkwell with a hinged lid; on the left are a match container and a small candle holder. The matches were used to light the candle, which in turn was used to melt wax for sealing the letter. c. 1885. 3" x 6", 3 3/4" tall. $800-1,200.

French bronze inkstand with a boy leaning on a bucket beside a well. The well has a hinged lid. On the far right is a hollow stump to hold a small candle. The open bucket on the left was used to hold matches. c. 1885. Marble base. 3 1/2" x 6 1/2", 4" tall. $800-1,200.

French Bronze inkstand on a marble base. The figure of a boy is standing in the middle of the composition. The inkwell is on the right side and opens on a hinge. The hollow log on the left holds matches. Next to the boy's outstretched hand is a small hole to hold a single match. There is a rabbit on the right side of the inkwell and another one is hiding in front. c. 1875. 4" x 7", 4 1/2" tall. $800-1,200.

A French bronze inkstand that has several unique features. A barefoot boy is sitting in the middle of the composition with a rabbit by his foot. The inkwell, with a hinged lid, is on the right. In the left rear is a picket fence enclosure to hold matches, and on the left front is a round hollow tree stump to hold a small candle. c. 1875. Marble base. 3 5/8" x 6 3/8", 4 1/2" tall. $800-1,200.

French bronze inkstand with a match holder. A barefoot boy is holding a rifle in his left hand and a match in his right. The hunter's hat and pouch are on top of the inkwell on the right side. The lid is hinged. The dog appears to be eager for the hunt. c. 1865. 3 1/2" x 5 1/4", 4" tall. $800-1,200.

Inkstand with the figure of a pageboy holding a letter in his left hand. The inkwells on either side are in a swirl pattern and the lids open on hinges. The base is made of onyx and stands on four pad feet. The channel across the front is for a pen. France, c. 1875. Brass. 4" x 7", 6 1/2" tall. $400-500.

A French bronze inkstand that has many interesting details. A hut by a millstream has a hinged roof. On the right side is a match container with a striker plate in the form of steps leading into the hut. On the left side is a candle holder, and in the back is a pen holder shaped like a split rail fence. The mill wheel is on a spindle and turns. c. 1875. Bronze. 3 1/4" x 5 1/4", 5" tall. (Inkstands with match holders are double collectibles.) $800-1,200.

A hut with a thatched roof that is hinged. A match holder is on the right side and the grooved steps leading into the hut are used to strike the match. Austria, c. 1880. Bronze on a wooden base. $300-500.

Bronze inkwell in the form of a thatched roof cottage. The roof is hinged. Two tree stumps have holes to hold a pen, a small candle, or matches. A dog and a bundle of grain sit outside the door. Europe, c. 1900. The base measures 4" in diameter, 4 1/2" tall. $300-400.

A basket woven hut with a thatched roof that is hinged. Bees and leaves add to the decoration. France, c. 1880. Bronze wire, alabaster base. 3 1/4" diameter, 5" tall. $300-400.

An Arab is standing on an Oriental rug holding a rod in his hand. The table on the right has a hinged top and is the inkwell. Resting on the table are a box and a teapot. Two urns, a book, and two rods are on the floor. Austria, c. 1910. Painted bronze. The Austrian black and white marble base has a pen channel across the front and measures 9 1/4" x 10 1/4". $1,800-2,500.

86

An Arab is sitting on an Oriental carpet smoking a long pipe. The inkwell with a hinged lid is on the right side. Austria, c. 1910. Painted white metal. 4 1/2" x 8 1/2", 4 3/4" tall. $500-600.

This unusual inkstand has a wind up music box that plays a tune when the hinged lid on the inkwell is raised. The scene is taken from *The Angelus Call to Prayer*, an oil painting by Jean Francois Millet that is in the Louvre in Paris. A peasant couple with heads bowed are heeding the church bells noon-time call to prayer. The oval base stands on four ball feet. France, c. 1895. Spelter. 2 1/4" x 5 1/2", 5 3/4" tall. $600-800.

The figure of a bearded man holding a child in his arms stands on the back of the inkstand. The inkwell in the center has a hinged lid. A plaque on the front reads "Sir Oratoires Joseph." Inscribed on the back "Made in France." c. 1895. Spelter. 4 1/4" x 5", 4 1/4" tall. $125-175.

Humorous inkstand with the figure of Diogenes, a Greek philosopher, holding a lantern looking for an honest man. The sign in the barrel reads "Defense Des Ordures De Cette Propriete." Loosely translated "Do not leave your refuse here." On the left is the inkwell with a hinged lid. Behind the out-stretched hand is a pen holder. France, c. 1885. Bronze. 3" x 4 1/2", 4" tall. $700-800.

The tam on the man's head is the hinged lid on the inkwell. There are holes on the top of the ears to hold pens. Inscribed on the base "Copyright Co Kato." America, c. 1900. Bronze patina over white metal. $175-250.

A seated man is wearing a turban and holding a pipe in his right hand. The shoulder level cover is removable. Porcelain. 3" x 3", 5" tall. $250-350.

A man's torso with short hair and a cleft chin. He is attired in a shirt, tie, vest, and jacket. The hinge is located in the back at the collar line. England, c. 1890. Bronze. 3 1/2" x 4 1/2", 6" tall. $600-850.

Inkwell in the form of an African man's head. The top of the head is hinged above the ears. c. 1890. Bronze. 3 1/2" x 4 1/2", 4" tall. $900-1,500.

A black man is sitting beside a basket inkwell. He is wrapped in a white robe and has a red and white striped turban on his head. Europe, c. 1910. $200-300.

An African man is playing a flute. A hinge is located in the back between the man's body and the bale. c. 1900. Painted white metal. 4 1/2" tall. $300-450.

A Moroccan wearing a red fez is sitting with crossed legs and has a pipe in his left hand. The man's body is hinged at the waist. France, c. 1900. Painted white metal. 2 1/2" x 2 1/2", 3 3/4" tall. $300-450.

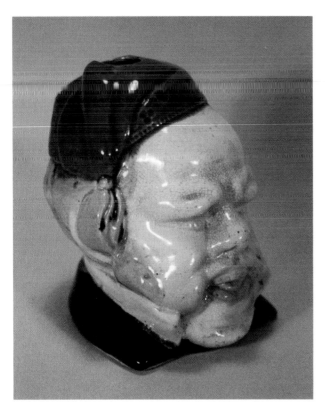

A man is wearing a cobalt blue sleeping cap with a tassel on the side. There is a quill hole in the top of his head. England, c. 1860. Highly glazed porcelain. 3" x 3", 3 3/4" tall. $350-450.

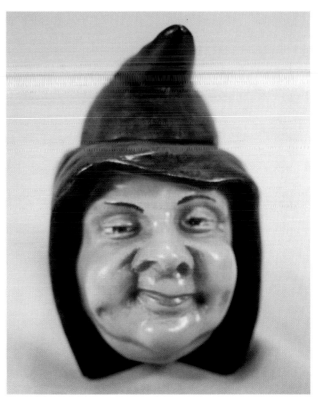

The head of a smiling monk is the inkwell. The peak of the cowl is hinged in the back. c. 1900. Painted white metal. 2" x 2 1/2", 3 3/4" tall. $125-200.

A bar connects two monks. One monk has a happy smile on his face; the other one is frowning. The peaks of their hoods are hinged and there are two inserts. America, c. 1900. White metal. 3" x 5 1/2", 3 3/4" tall. $225-300.

Whimsical inkwell displaying a man in a top hat with his hands attached to the lid of a barrel. When the coat tails are pushed down, his hands lift the lid to expose a glass bottle inside. Made in Italy, c. 1900. 7 3/4" tall. $200-300.

In 1880, Carlo Lorenzini, an Italian author, wrote a story about Pinocchio, a wooden puppet boy whose nose grew longer when he told a lie. The tale was an instant success and has been delighting children ever since. Pinocchio's hat is the hinged lid. The painted wooden figure is attached to a wooden base. Europe, c. 1920. 3 7/8" in diameter, 4" tall. $200-300.

White porcelain bust of Richard Wagner, a German musician-composer (1813-1883). There is a hole on his shoulder for a quill or pen. Germany, c. 1900. 2 1/2" x 2 1/2", 4" tall. $200-300.

The last Tuesday before Lent is Mardi Gras, a day of merry making and carnival in New Orleans. The celebration marks the last day before the commencement of forty days of solemnity. This inkstand depicts Rex, the king of carnival, wearing a crown that is hinged in the back. The date 1911 is embossed on the right side. America. Brass. 7" wide. $200-300.

A Spanish soldier is standing on a pedestal holding a long oar that is a letter opener. The figure of the soldier is a seal that can be removed to close an envelope with hot wax. The roped bale has a hinged lid and opens to expose the inkwell. There is a barrel behind the bale (not visible in this picture) that holds pen nibs. An anchor completes the decoration. Europe, c. 1890. Bronze on a marble base. 3 1/2" x 5", 8 1/2" tall. $400-600.

An eagle with spread wings forms a tray that displays a rising sun and a portrait of Napoleon Bonaparte. A square glass inkwell with a loose lid is set in a laurel leaf wreath. France. The base is c. 1880; the inkwell is c. 1920 and not original. Copper coated iron. 5 3/8" x 6 1/4", 2 1/8" tall. $150-200.

Napoleon is leaning on his rifle beside a wounded soldier. The tree stump inkwell has a hinged lid. France, c. 1875. Bronze. 3 3/4" x 6 1/4" (oddly shaped base), 3 3/4" tall. $300-400.

Salt glaze pottery in the shape of a man's head. His open mouth is the inkwell and there is a hole in the forehead to hold a quill. England, c. 1830. 2 3/8" x 2 1/2", 1 7/8" tall. $200-300.

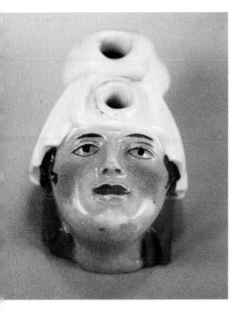  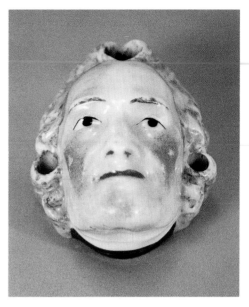

The head of a woman with dark hair and rosy cheeks. She is wearing a white cap and red collar. There are two holes in her hat: one for ink, and the other for a quill or pen. France, c. 1840. Porcelain. $250-350.

Inkwell in the shape of a man's face. In his cap are two holes; one is a dip hole for ink and the smaller one is to hold a quill. Rockingham pottery. England, c. 1840. 2 1/4" x 3", 1 1/3" tall. $200-300.

Porcelain figure of an English gentleman with an inkwell in the top of his head. There are quill holes in the coils of his hair. England, c. 1830. 2 1/2" x 2 3/8", 1 1/2" tall. $300-450.

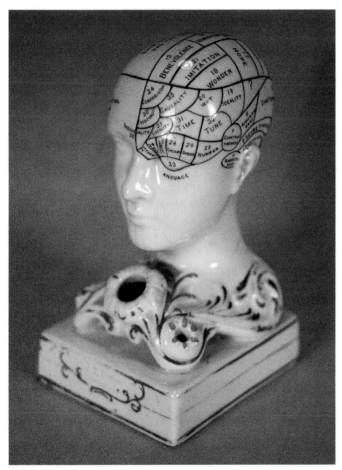 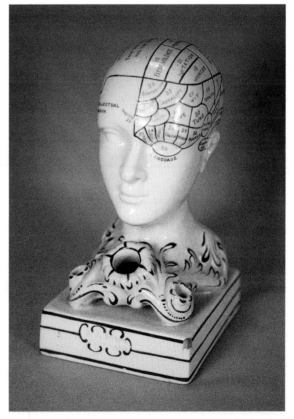

Franze Joseph Gall, an anatomist in Germany in the eighteenth century, attempted to divine individual character and intellect by the shape of the skull. Phrenology was used to map the functional areas of the brain. Various manufacturers in Vermont are credited with making these inkwells. Some in granite ware or porcelain are attributed to Bennington Pottery. The inkwell is the hole in the center and there are small quill holes on each side. America, c. 1860. White porcelain trimmed in gold. 3" x 3 1/4", 5 1/2" tall. $1,800-2,500.

Phrenology showing the areas of the brain as mapped in gold on the skull. The inkwell is in the center of the base and there are small quill holes on the sides. Trimmed in blue. Vermont, America, c. 1860s. 3" x 3 1/4", 5 1/2" tall. $1,800-2,500.

The symbol of a skull and cross bones has been used for centuries to alert people to danger. This emblem used on a bottle of chemicals warns that the contents are poisonous. Pirates used the skull and cross bones on their flags to strike fear in the hearts of their victims. Hinged lid. c. 1890. Brass. 3" x 3", 2 1/4" tall. $400-500.

A skull wearing a crown is sitting on a book. The top is hinged. On the lower left-hand side is a pen holder. c. 1900. Brass. 2" x 2 1/2", 2" tall. $200-300.

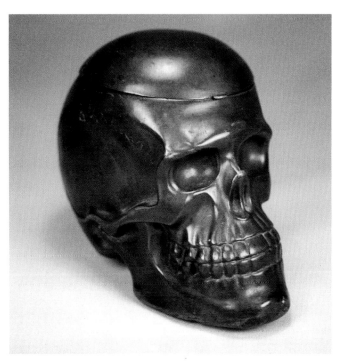

A large bronze skull with a hinge. America, c. 1870. 3" x 4 1/2", 3 1/4" tall. $700-900.

A grotesque mask covers the hinged lid. Imprinted "Austria" on the bottom. c. 1900. Bronze. 3 1/2" in diameter, 2 3/4" tall. $200-300.

An inkwell with a hinged lid is located behind the smokestack on top of the train. A pen rack is mounted on the base and there is a round open compartment for pen nibs or sealing wafers. England, c. 1880. Mahogany and brass. 4 1/2" x 8". $350-450.

The bizarre face of a laughing man is framed with a scalloped collar that terminates in round balls. The lid is hinged at the top. Europe, c. 1900. Brass. 4 1/8" x 5 3/4", 3 1/4" tall. $200-300.

Hand-carved wooden inkstand with a saddle on top of a horse blanket. The inkwell with a domed lid is behind the saddle. England, c. 1885. 5" x 7 3/4". $700-950.

A four-wheeled wooden cart has two clear crystal inkwells mounted between posts on the flat bed. The faceted lids are hinged with brass mounts. A brass railing is attached to posts on the corners. England, c. 1900. Olivewood. 3 1/2" x 5 5/8", 3 1/2" tall. $600-700.

A round inkwell, a rectangular two compartment stamp box, and a stack of round shot surround a pistol on a wooden base. The tops of the hinged lids are littered with shot and casings. England, c. 1885. Wood and brass. 6" x 8 1/4", 2 1/4" tall. $650-950.

A World War I soldier is manning a machine gun. There is an ammunition box at his feet and cartridge belts on the side. The inkwell is in the mound on the left side. France, c. 1916. Spelter. 3 7/8" x 6 1/4" x 3 7/8". $300-400.

Four crossed muskets form a pen rack. A crystal inkwell with a loose lid is fitted in the base. Marked on the bottom "James W. Tufts. Quadruple Plate #2864." Boston, Massachusetts. America, c. 1885. 3 5/8" x 3 5/8", 3" tall. $300-400.

A World War I French soldier is manning a machine gun with a box of ammunition by his side. "VIMY" is engraved on the front. The inkwell on the left side has a hinged lid. Marked "Richer. Paris. Deposé. Made in France." c. 1916. Spelter. 4 3/4" x 6 3/4", 3 3/4" tall. $450-550.

Inkstand in the shape of a battleship with guns fore and aft. The top lifts off to expose two oddly shaped glass inserts. Germany, c. 1916. White metal.
2 1/4" x 7 3/4", 3 1/4" tall. $300-400.

View of the battleship with the top removed to reveal the oddly shaped inserts.

Helmets are a form of armor that protect the head. Military helmets have been worn since ancient times. This helmet shaped inkwell is hinged in the back. France, c. 1885. White metal.
3 1/2" x 3/1/2", 3 1/2" tall. $200-250.

Unique helmet with a mother-of-pearl crown. France, c. 1860. Bronze on an alabaster base. 3 3/4" x 5", 6 1/4" tall. $500-700.

A German World War I helmet with the Prussian eagle on the front. The spike on the top is the removable lid. Marked "Gesetzlich Geschutzt" (meaning "Legally Protected"). Germany, c. early 1900s. 2" x 3", 3 1/4" tall. $200-250.

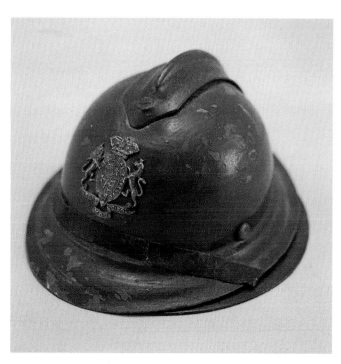

A World War I helmet with a crest on the front. The helmet is hinged in the back. France, c. 1916. Painted aluminum. 3" x 3 1/2", 1 1/2" tall. $200-250.

German helmet. The spike on the disc-shaped hinged lid is the finial. Engraved "Mit Gott Fur Und Koenig Vaterlad. F R." ("with God for king and fatherland"). c. 1900. Silverplated. 2 1/2" x 3 1/2", 3 1/2" tall. $300-500.

Japanese warrior's helmet with a red tassel. The lid is hinged in the back and opens to reveal two unusual half-moon shaped glass inserts. c. 1900. White metal with gold trim. 3 1/2" in diameter, 2 1/2" tall. $400-600.

A Japanese warrior's helmet with the figure of a dragon on the front. The top of the helmet is hinged in the back. c. 1900. Silverplated white metal. 3 3/4" x 4", 2 1/2" tall. $400-600.

Basket shaped inkwell constructed of brass wire. The hinged lid is made of copper. Various seashells decorate the top. Japan, c. 1900. $200-250.

"The Victory" under full sail. The inkwell has a cover that is chained to the ship. The oblong channel on the left is for the pen nibs. England, c. 1920. $75-125.

In Victorian days, the lustrous inner lining of the pearl oyster was often used on articles such as card cases, knife handles, letter openers, and numerous other items including pens and inkwells. A crystal inkwell is seated in a pierced brass frame on the deck of a mother-of-pearl ship. The faceted lid has a brass collar and is hinged. The rack across the front holds a pen. America, c. 1880. Mother-of-pearl base, ship, and sails, with brass riggings. 3 1/2" x 5 3/4", 7 1/2" tall. $500-700.

A weighted tether for a horse. The hinged lid with a ring has a safety catch. Inscribed #7 in an oval on the front. Covered with an engraved design. c. 1870. Brass. 3 3/4" in diameter, 5 1/2" tall (including the ring). $350-450.

A mother-of-pearl inkstand that is shaped like a boat. A crystal inkwell with a faceted hinged lid sits on the deck. A brass railing runs around the top of the boat. Displayed in the pen rack in the back is a mother-of-pearl pen with "Los Angeles, Cal." inscribed on it. America, c. 1900. 2 1/4" x 7 1/2", 3" tall. $175-250.

A heavy brass scale weight with a twist-off cover. Marked "SCH 6" on the top. England, c. 1860. 2 3/8" in diameter, 4 1/4" tall. $250-350.

Glass and brass inkwell that resembles a light bulb in a socket. The bulb reservoir unscrews for filling and there is a dip hole in front. The round base has a groove for a pen. America, c. 1870. 3 1/2" in diameter, 4 1/2" tall. $125-175.

Usual shaped inkstand, a ferryboat perhaps. A square pressed glass inkwell is in the back; the metal lid is hinged. A pen wiper brush and a pen rest are in the front. c. 1880. Silverplate. 2 1/2" x 5". $250-300.

Brass and glass inkwell in the shape of a lantern with handles. America, c. 1880. 2 1/2" in diameter, 3 1/2" tall. $125-175.

Glass with a loose brass cover. On one side of the top is an oval saucer to prevent overflow when the attached glass funnel is used to fill the inkwell. America, c. 1890. 3 5/8" in diameter, 3 1/4" tall. $150-250.

Small curling stone with a hinged lid. The message on the side reads "Souvenir of Aberdeen." Scotland, c. 1920. 2 1/2" in diameter, 2 1/8" tall. $125-200.

Curling stone with a hinged lid. "Curling" is a game especially associated with Scotland where it has been played since the sixth century. It is similar to lawn bowling but is played on ice. A unique feature of the game is that teammates use a broom to sweep the ice in the path of the oncoming stone, which gives the stone a longer, smoother slide. Scotland, c. 1920. 3" in diameter, 2 1/2" tall. $150-225.

A three-wheeled leather bath cart with an ivory handle. Invalids at a spa or the beach used this type of cart. The clear crystal inkwell has a faceted lid that is hinged with a brass collar. There are three ledges to hold pens. A cloth pen wiper is tucked behind the inkwell inside the cart. England, c. 1870. Brass wheels. 2" x 4 1/4", 3 3/8" tall. $500-700.

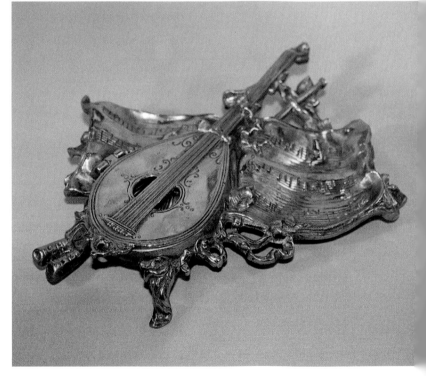

A mandolin is resting on a bed of sheet music and ribbons. The top of the instrument is hinged on the backside and opens to expose a porcelain insert and a crescent shaped compartment to hold pen nibs. England, c. 1890. Brass. 7" x 8 3/4". $450-650.

Rosewood grand piano with an inlaid silver design on the top. The lid is hinged and opens to reveal two porcelain inserts. Supported on three legs. Under the lid is the inscription "Bur Srinnerung. 1. Juni 1868." Germany. 8" x 15", 4 1/4" tall. $1,000-1,400.

Inkhorn that has a hinged lid with a finial. Encircled with a brass band that terminates in a tassel. Although a bit large for a traveler's inkwell, it has an interior-locking device and could have been converted for the desk. Sweden, c. 1880. Horn with brass mounts. 2 1/2" x 3 1/2", 4 1/2" tall. $300-400.

A floral cut crystal inkwell with a hinged lid sits in a leaf shaped frame. The well is fastened to a pair of horns that have a curled pen holder at the top. America, c. 1900. The frame, lid, pen holder, and caps on the ends of the horns are made of brass. The inkwell is 3 3/8" x 3 3/8", 2 3/4" tall. Horns are 9 1/4" wide, 11" tall. $400-550.

A crystal inkwell is fitted into a base that is attached to a long antler. America, c. 1890. Brass hinged lid and base. 5" x 9 1/2", 2 3/4" tall. $300-400.

A clear glass inkwell sits between a pair of horns that form a pen rack. The brass lid is hinged. Germany, c. 1880. Black wooden base. 5 3/4" in diameter, 5" tall. $200-300.

Inkwell in the shape of a heater. There is a round hinged porthole at the top that opens to give access to the inkwell. On the back is a hinged trap door that opens for filling the reservoir with ink. Embossed on the back "Ancne Mon Godin." France, c. 1890. Iron. $250-350.

Inkstand in the shape of a footed iron stove. On top of the hinged lid is written "CHAUFFETTE," and on the front of the door "GODIN." Stamped on the bottom and the inside is number 15. France, c. 1900. 3" x 5", 4 1/2" tall. $250-350.

Urn on a pedestal. The base has a marble center to match the marble on the inkwell. The square base is footed and there are handles on the sides. The hinged lid has a finial. England, c. 1900. 4" x 5 1/4", 4 3/8" tall. $250-350.

Inkwell in the form of a fire hydrant and hose. The lid is hinged. Inscribed on top "Insurance From Loss F A 1817" and around the base "Fire Association Of Philadelphia 1814-1917." America. White metal. $175-225.

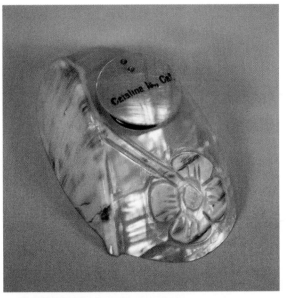

Souvenir from Catalina Island, California. Carved mother-of-pearl with a brass hinge. America, c. 1925. 2 3/4" x 4 1/2", 1 1/8" tall. $50-75.

A brown and gold tin box advertising "Hall's "STATE" Toffee. Sole Manufacturers, Hall's Bros. Ltd. Whitefield, Manchester." The hinged lid opens to expose two red pottery inkwells and a long channel for pens. England, c. 1928. 5 1/4" x 9 3/8", 5 1/4" tall. $200-300.

Open view of the Hall's inkstand showing the advertisement under the lid.

Souvenir in the shape of a fish. A scenic picture of a harbor with "Southend on Sea." written underneath. The two inkwells have hinged lids with brass collars. England, c. 1900. Glass. 1 1/2" x 5 1/2", 2" tall. $80-120.

Souvenir in the shape of a glass boat. A scenic picture is framed in the back with "Bleistein" written underneath. The two inkwells have hinged lids with brass collars. Germany, c. 1900. 1 1/2" x 5 1/2", 2" tall. $80-120.

Two souvenir inkstands with matching pens on pen racks. The one on the left has a single swirl glass inkwell. Framed in the back is the picture of a child. On the right is a double inkstand with two swirl glass inkwells with a picture in the back of "Margate." England, c. 1890. Carved bone. Left: 1 1/8" x 1 3/4", 2 3/8" tall. Right: 1 1/2" x 2 3/4", 2 1/2" tall. $150-200 each.

Inkstand that displays objects that represent art, science, and engineering. The lid is hinged. America, c. 1880. Brass. 8 3/4" in diameter. $600-700.

A craftsman with a hammer by his side is blowing on coals in an open oven. The inkwell with a hinged lid is located on top of the stove. There is a pen stand behind the man. India, c. 1920. Brass. 3" x 4 3/4", 3" tall. $175-225.

Souvenir from France with a replica of Notre Dame in the center. The two inkwells have Paris scenes on top of their hinged lids. The tray across the front is for pens. c. 1900. Painted white metal. 3 1/2" x 8", 3 1/2" tall. $100-150.

Art Deco style souvenir with a single inkwell in front that has a hinged lid. A pen tray is across the middle. Written on the base of the building on the left is "Arc De Triomphe," and on the right is "Le Sacré Coeur. Paris." Inscribed on the bottom is "R C 358. Made In France. Deposé." c. 1930. 4 3/4" x 5 1/8", 3 1/8" tall. $95-145.

A souvenir from Paris. On the top of the inkwell on the right is a picture of the Eiffel Tower, and on the left is a picture of the Arc De Triomphe. The figure of a prancing horse with a rider carrying a banner is on the top. Across the lower front are ledges to hold pens. France, c. 1900. Cast brass. 3 3/4" x 7 1/4", 6 3/8" tall. $400-500.

Souvenir from France. Inscribed on the front of the building on the right is "SC MONTARE PARIS," on the left is "PARIS ARC DE TRIOMPHE," on the back of each building is "LL PARIS," and on the bottom is "14 LL MADE IN FRANCE." The Eiffel Tower is on top of the hinged lid. c. 1920. Spelter. 3 3/4" x 5 5/8", 4 1/4". $125-175.

Replica of the Capital in Washington, D. C. The top half of the inkstand is hinged across the back and opens to expose two milk glass inserts. Stands on four foliated feet. America, c. 1900. White metal with a brass finish. 3" x 4 3/8", 3 1/2" tall. $150-200.

Inscribed on the top is "Expo Colonial 1931." On the bottom is "Temple D' Angkor. Made in France. 200." Signed "Ouveh." On the church is #2137. White metal. 2 3/4" x 6 1/4", 3" tall. $150-190.

Souvenir from Paris. A footed Art Deco style inkstand displays a replica of the Reims Cathedral on top. The Arc De Triomphe is depicted on the hinged cover. On the bottom is "Deposé 201." France, c. 1920. White metal. 4 5/8" x 5 1/4", 3 1/4" tall. $100-150.

Armistice train commemorating the end of World War I. Inscribed under the train "Wagon Du Marechal Focr Oufut Signel Armistice Le in Novembre 1918." The single inkwell in front has a hinged lid. Cut out design on the base with a flower and leaves. Deep pen tray. France. White metal. 4" x 5". $200-300.

Souvenir from Canada displaying "The Basilica, The Parliament Bldg., General View of Quebec." The lid is hinged across the back and opens to expose two inserts. A pair of Canadian geese are wrapped around the sides. Brass. 3 1/2" x 5". $175-225.

Early American blown three-mold black glass inkwell. Coventry Corners. c. 1840. 2 1/4" in diameter, 1 1/2" tall. $125-200.

Black ware inkwell with a funnel hole. There are four holes in the top to accommodate quills or pens. Probably Wedgewood. England. Basalt. 3 1/2" in diameter, 2 5/8" tall. $200-300.

Red ware inkstand with three compartments: an inkwell, a sander, and a small seal container in the back. America. 3" x 4 1/2", 2 1/4" tall. $150-250.

Three pressed glass inkwells with loose lids are sitting on a cast iron base. A pen holder for two pens is on the front. America, c. 1885. 5 1/2" x 9 1/2", 4" tall. $250-400.

Opaque glass boat with an inkwell in the center. The brass hinged lid has an anchor and chain design on the top. America, c. 1910. $150-200.

Black bakelite with a hinged lid. A gold lizard on the top is the only decoration. America, c. 1920. 3 1/4" x 5 1/2". $95-150.

Black lacquered double inkstand with a handle. The hinged lids are covered with palm trees and flowers. An applied scroll decoration is on the front. c. 1920. 3 1/2" x 6 3/8", 3 1/2" tall. $150-250.

Black glass inkwell with pen channels on all four sides. The matching lid is hinged. America, c. 1925. $75-125.

Blue and white spatter inkwell. The lid is loose and there are two quill holes on the corners. Marked "Semovices Limoges." France, c. 1885. 3 1/4" x 3 1/4", 2 1/4" tall. $200-300.

A saucer shaped base with an inkwell in the center. The matching domed lid is hinged with a brass collar. Decorated with blue flowers, leaves, and birds. England, c. 1890. 3 3/4" diameter, 2 1/4" tall. $125-200.

Above: Arts & Crafts handcrafted inkwell with a hinged lid. Roycroft hallmark. America, c. 1910. Bronze. 4" in diameter, 2" tall. $250-350.

Left: Orange porcelain inkwell with pink flowers on a white background. The brass cover has a hinged lid and two quill holes. Marked "Aladin Made in France." Limoges. c. 1895. 3" in diameter (around the top), 2 1/8" tall. $200-250.

Arts & Crafts inkwell with a hunter holding a fox above his head to keep it out of the reach of the hounds below him. Written on the left-hand side in embossed letters "The Death." There are additional hunting scenes on the sides. The hinged lid has an open arch finial. England, c. 1910. Brass. 5" x 5 3/8", 6 3/4" tall. $650-850.

Arts & Crafts hand wrought copper and brass double inkstand. The two pyramid shaped hinged lids have ball finials. America, c. 1890. 5" x 7 1/4", 4 1/2" tall. $700-900.

This inkwell is part of an eight-piece desk set that was made for Marshal Fields department store. Inscribed "M.F. & Co. Athenian. Made in Milwaukee, Wisc." The top of the hinged lid is embossed with horsemen. Inscribed around the rim "Self Closing Inkstand Co. Pat. Apr. 1, 06. Aug. 25, 04. Jan. 18, 07. Sept. 1, 14." America. Bronze. 5 1/2" in diameter, 4" tall. $300-500.

Arts & Crafts octagonal shaped inkwell with a sterling silver overlay of flowers and vines. Heintz Art Metal (maker). c. 1910. Sterling on bronze. 5 1/8", 3" tall. $350-500.

Hammered copper inkwell. American Arts & Crafts Movement (1895-1920). The hinged lid has three leaves decorating the top and opens to expose a green glass insert.
3" x 3", 1 5/8" tall. $95-150.

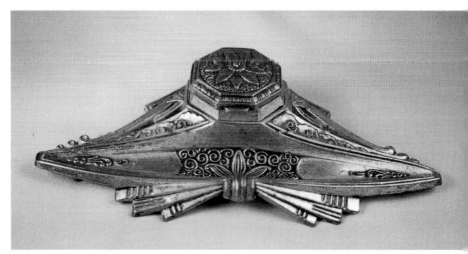

Art Deco inkwell with a pen tray across the front. The hinged lid has a flower on top. France, c. 1925. 3" x 5 1/2", 1" tall. $100-135.

Arts & Crafts hand-hammered copper. Marks on the bottom identify this as a Roycroft piece. The sliding cover on the inkwell reads "Patent Pending. Grand Rapids, Mich." America, c. 1910. 2 3/4" x 2 3/4". $200-300.

Embossed chrysanthemums in a circle are displayed around the sides and on top. The lid is hinged. Japan, c. 1890. 2 3/8" x 2 3/8", 3" tall. $250-325.

Art Deco hexagonal shaped with each section filled with a floral design. The lid is hinged. America, c. 1925. White metal. 2" in diameter, 2 1/2" tall. $50-75.

Inscribed around the middle "Matthew's- Marvcs-Lycas-Johannes." The lid is hinged. England, c. 1925. White metal. 3 1/2" in diameter, 4" tall. $100-150.

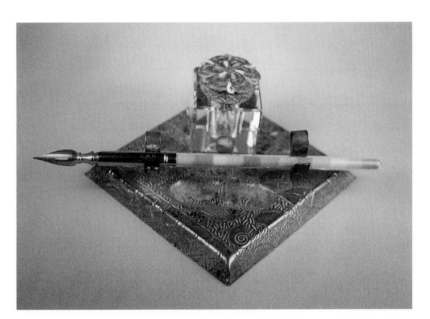

The crystal inkwell has a removable lid with enameled flowers decorating the top. A pen holder is across the triangular shaped base. Rolled brass with a leafy design over the entire surface. America, c. 1920. 4" x 4", 2" tall. $75-125.

Middle Eastern scribe's inkwell. A scribe was an educated person who was hired to write documents and letters for people who could not read or write. Dried ink was put in the small inkwell at the top and then the scribe would use water or sometimes saliva to liquefy the ink. The lid is hinged. c. 1850. Brass. 1 1/2" x 3". $150-250.

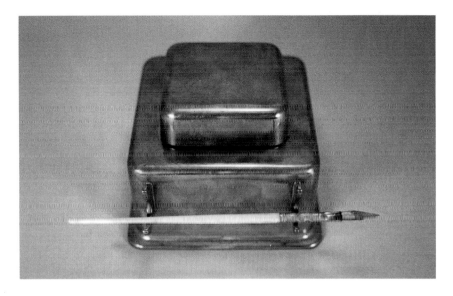

Square inkwell with a hinged lid. A pen rack is on the front. There is a mark for K. & O. Metal Novelties Co. America, c. 1930. Brass. 4 1/4" x 4 1/4", 3" tall. $150-225.

Oriental casket with a pen rest across the front. The lid is hinged and opens to expose two porcelain inserts and a pen wiper brush. c. 1910. White metal. 2 3/4" x 4 3/4". $150-295.

The pen rack in the back holds four pens. The inkwell has a funnel dip hole and sits on a scalloped edged tray. The entire composition is covered with a floral and leaf design. India, c. 1890. Copper. 4 1/2" x 6 1/2", 3 1/2" tall. $195-225.

Art Nouveau casket covered with embossed flowers. The hinged lid opens to reveal two milk glass inserts. c. 1910. Nickel-plated. 4" x 5 1/8". $200-250.

Above: Oriental motif with an ornate floral and scroll design on the top. When the hinged lid is opened, a beautiful garden scene is pictured on the inside of the lid. Fitted with two white porcelain inserts and a nib cleaning brush. A pen rack is on the front. Hong Kong, China, c. 1900. Copper and brass. 3 1/2" x 4 1/2", 1 3/4" tall. $175-295.

Left: Open view showing the scene under the lid.

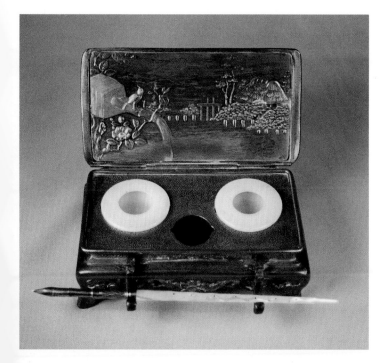

A copper and brass kettle stands on three legs. On the front is an applied image of Reims Cathedral with a fleur-de-lis on each side. The removable lid has a ball finial. France, c. 1900. $125-150.

Small inkwell in the shape of an Oriental teapot with a handle. The lid is attached by a chain. On the front is a rickshaw and on the back a fly. Japan, c. 1890. Brass. 2" in diameter, 1 3/4" tall. $200-300.

Double inkstand in an Oriental design. Beautifully pierced backplate with a pen rack to hold four pens. The top is covered with a water dragon and slides back to expose two milk glass inserts and a small round brush for cleaning the pen nibs. Stands on four scroll feet. America, c. 1900. White metal with a copper patina. 3" x 5", 4 1/2" tall. $200-300.

Open view of the Oriental inkstand showing the sliding lid.

Oriental writing box with a bronze overlay on wood. Wading cranes and irises decorate the top. The hinged lid opens to reveal two opalescent inserts, and a nib cleaning brush. When the hinged lid is open, the side supports form a pen rack. Japan, c. 1900. 2 1/2" x 4 1/2", 2 1/8" tall. $300-400.

Oriental blue and white porcelain with a flying crane on the front. The loose lid is covered with flowers and has a ball finial. 2 7/8" in diameter, 2 1/2" tall. $300-500.

Oriental style with an open grillwork backplate decorated with irises and leaves. The pen rack down the sides holds four pens. The sliding lid is covered with a water dragon and opens to expose two milk glass inserts and a small round brush for cleaning the pen nibs. Scroll feet. America, c. 1900. White metal with a copper patina. 3" x 5", 4 1/2" tall. $200-300.

Hexagonal shaped Peking enamel inkwell. The yellow background is covered with a pink, blue, and green floral design with scenic pictures on the front and back. 3" in diameter, 2 1/4" tall. $300-500.

Hand painted Nippon with a floral design and a stylized rooster in an archway. Rising sun green mark on the bottom. Japan, c. 1900. 4" x 4", 4" tall. $175-250.

Porcelain blue and white floral bell-shaped inkwell with brass mounts. The hinged lid is closed with a decorative catch. There is an English registry mark for 1883. 3 1/2" in diameter, 4" tall. $350-500.

Square footed inkstand with a reticulated border. A blue and white porcelain inkwell is fitted in a frame on the top. The mushroom shaped lid is hinged. Inscribed "E. G. Webster & Son. N. Y." America, c. 1890. 3 3/4" x 3 3/4", 3 1/2" tall. $300-400.

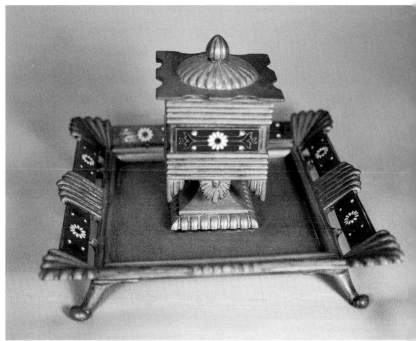

Footed inkstand with a square urn on a pedestal. The lid is hinged. The base has an open railing with red, white, and blue enameled flowers. France, c. 1900. Brass. 6" x 6", 4 1/4" tall. $600-800.

A cloisonné inkwell that has a hinged lid stands on an alabaster base with a champlevé border. A pen rack is on the front. France, c. 1880. 9 3/8" square. $300-500.

Tiny mosaic inkwell with a hinged brass collar. c. 1915. 2 1/8" in diameter, 1" tall. $125-175.

Champlevé is a process that consists of cutting cells or channels in a metal plate. The raised metal lines form the design and are then filled with colorful pulverized vitreous enamel. This urn shaped inkwell sits on a saucer base with irregular formed feet. The hinged lid has a ball finial. Europe, c. 1900. 3 5/8" x 4 1/4", 2 1/2" tall. $250-350.

Round blue and gold ceramic with a hinged brass lid. England, c. 1900. 2 1/4" in diameter, 1 1/2" tall. $150-300.

Porcelain and brass inkstand with two decorative scroll handles. The lid is hinged and has an ornate finial. A burgundy red and gold design on a white background adds to the beauty of this piece. France, c. 1880. 5 1/2" x 5 1/2", 5" tall. $600-800.

This urn shaped inkstand has a hinged lid with a pineapple finial. The composition is supported on three bronze legs surmounted by winged cherubs and terminating in paw feet. A colorful champlevé band encircles the mid section. France, c. 1885. Green onyx. 3 1/4" in diameter, 6 1/2" tall. $500-700.

French enamel champlevé inkstand on an onyx base. The single inkwell has a hinged lid. Supported on six pad feet. c. 1890. 6 3/4" x 9", 3 1/4" tall. $400-600.

Delicate Oriental scenes in red and black, enhanced with gold. The lid is loose. Marked on the bottom "Carlton Ware. Made In England. #2880W." c. 1920. 3 3/4" x 7 3/4". $125-175.

Emerald green art pottery with a speckle glaze. Fitted with two inkwells that have loose lids with ball finials. There are two holes on top to accommodate pens. England, c. 1900. 3" x 4 3/4". $150-250.

Highly decorated octagonal inkstand. There is a stamp box in the rear, a pen rack, a candle holder in the middle, two holes for seals, and two inkwells in front. The inkwells have removable lids with gold finials. Covered in a lovely Oriental floral pattern. A mark on the bottom indicates this was made by Samson in Paris, France. c. 1870. Porcelain. 7 1/8" x 7 1/8", 2 3/4" tall. $600-800.

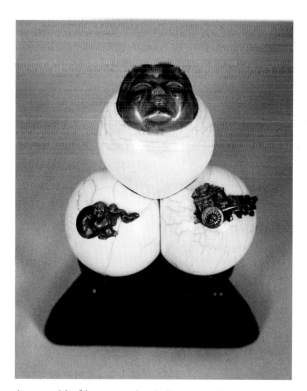

A pyramid of ivory snooker balls are held together with ivory pegs. A cherub's face decorates the hinged lid on the top ball. Characters on the sides include a samurai warrior, an ox cart, and a bee on the back. The composition rests on bamboo style legs on a triangular ebony base. Bronze mounts. Europe, c. 1860. 3 3/4" x 3 3/4", 5 1/4" tall. $700-900.

This inkstand has many interesting features: two inkwells with loose lids, a candle holder, two seal holes with one seal, two quill holes in front, and a wax wafer box in the center. This style inkstand always included a matching seal. Cantonese style decorations in a yellow, green, pink, and red. Marked on the bottom "B. C." in the center of a six-pointed star and "B. R. 8. Limoges, France." c. 1850. Porcelain with brass mounts. 5" in diameter, 3 3/4" tall. $1,000-1,400.

Orange and black porcelain with a blue and white bug on the loose cover. There are two quill holes on the top. Marked "Bastard Park. Made In France." c. 1925. $250-300.

Green, orange, and black flowers on a white background. Signed "Margeril." Marked "Bastard Paris. 16/1400. Made in France." c. 1925. 3 1/8" in diameter, 1 1/2" tall. $250-300.

Triangular shaped with a loose cover and a ball finial. Decorated with pink and yellow flowers with blue and green leaves. Signed on the bottom "Georg Schmider." Austrian faience, c. 1920. 4 1/2" x 4 1/2", 3 1/4" tall. $125-175.

Heart shaped with hand painted roses on the loose cover. Inscribed on the bottom "Elize. S. Head M. Limoges, France, c. 1910." Porcelain. 2 1/4" x 2 3/4", 2" tall. $175-250.

Brightly decorated Gouda pottery. A small paper label says this item was made in "Zuid, Holland." Before 1900. 3 1/8" in diameter, 3 3/4" tall. $250-350.

Square porcelain Art Deco inkwell marked "Kisc or Kiss." Yellow, black, and white with orange flowers. France, c. 1920. 2 1/2" x 2 1/2", 1 3/4" tall. $95-150.

Small footed dark blue cloisonné inkstand. The inkwell a has loose lid. The curved tray in front is a pen rest. Japan, c. 1895. 3" x 5", 2 1/2" tall. $250-325.

Pretty porcelain inkstand embellished with delicate flowers and draped wreaths. The brass top has an inkwell with a hinged lid, two quill holes, and two larger holes for seals. Missing the matching seal. Marked on the bottom "G.D.A. France. Gerald D. Abbot, Limoges." c. 1885. 3 1/4" in diameter, 5 1/2" tall. $250-400.

Blue and white porcelain with a brass cover. The inkwell in the center has a candle holder on the top and is surrounded by four holes, two for quills or pens, and two for seals. Missing the matching seal. A framed farm scene is on the front and a different scene is pictured on the back. Inscribed on the bottom "France M. De M. Limoges." c. 1895. The base is 2 1/2" in diameter and the top is 3 3/4" in diameter. 4 3/8" tall. $250-400.

Dark blue porcelain with pink roses and blue flowers in a white oval on the front. Small gold and white flowers are scattered over the entire stand. The two seal holes in the back are missing their brass rims. The matching seal in front of the inkstand fits in one of the holes. There are two quill holes in front. The hinged lid has a gold finial. Two blue crossed swords are on the bottom with the inscription "Made in France." c. 1890. 3 3/4" x 7 1/4", 4" tall. $400-500.

Beautiful porcelain basket with a handle that is entwined with roses. The brass top has an inkwell on the right side with a hinged lid. On the left side are three holes: one for a quill, one for a matching porcelain seal, and one for another seal. On the bottom (inside a triangle) is a "P" and "Made in France." c. 1880. 3 1/8" x 6", 6 1/4" tall. $800-1,200.

Porcelain with a brass top. The inkwell has a hinged lid. There are two quill holes, two seal holes with one matching seal, and a candle holder. The white background is beautifully painted with pink flowers and blue and yellow trim. Gold rings on the sides add to the decoration. Impressed on the bottom are a stylized "g" and a painted red mark. France, c. 1890. Bottom: 2 1/8" in diameter; brass top: 3 1/2" in diameter. 3 1/2" tall. $300-450.

Porcelain urn with a brass cover. The inkwell in the center has a hinged lid with a decorative finial. There are five holes in the top: one to hold a matching seal, one for another seal or sealing wax, one for a small candle, and two for quills or pens. An Oriental garden scene is on a black background around the sides. Marked on the bottom "France, M Dem Limoges." c. 1890. 3" in diameter, 5 1/2" tall. $300-450.

White porcelain inkstand decorated with dainty pink and blue flowers and accentuated with a gold band. The brass top has many functions. There are two inkwells that have hinged lids with finials: one matching porcelain seal in a hole, a hole for a second seal, two quill holes, a hole in the rear for a small candle, a stamp box in the center with a hinged lid, and an open compartment in front for pen nibs. Marked on the bottom "Alachin, France." c. 1890. 5 3/8" x 7 1/4". $500-700.

Dainty roses and blue ribbons decorate this inkstand and matching candle holder. The inkwell is on a saucer shaped base and has a loose lid with a finial. The candle holder has a handle. Marked in a wreathe on the bottom of the inkwell "Austria. Stouen. R. Briggs Co. Boston," and on the candle holder "Austria 1597." c. 1900. $300-400.

Porcelain with four holes on the top: two for pens and two for seals. The round lid has a knob finial and is removable. White background with green stripes. Metallic gold crickets are painted on the top and sides. Marked on the bottom "Made For I. Magnin & Co. N. Made In France." c. 1910. 3 1/2" x 6 1/2", 3" tall. $300-400.

Gondola shaped stand with a red border and floral decorations. The top has an inkwell with a loose lid, two seal holes with one seal, and two quill holes. Marked "Leares France 11019." An initial and 03 are on the bottom. Porcelain. 3 1/4" x 7 3/4", 3 1/4" tall. $300-500.

Porcelain with a ring handle and five quill holes. Decorated with a band of dark red and green flowers and trimmed in gold. On the bottom is a Bloor Derby mark (now Royal Daulton). Derby, England. 4 1/8" in diameter, 2" tall. $500-800.

Deep recessed saucer with a central inkwell and three quill holes. Pink, blue, and green flowers on a black background. A yellow band around the top adds to the decoration. The lid is removable. France, c. 1900. 3 3/4" in diameter, 2 3/8" tall. $200-300.

Porcelain inkwell with hand painted flowers enhanced with a gold trim. There are cups on each side and three tubes applied to the front for quills. England, c. 1820. 5" in diameter, 2 1/2" tall. $800-1,000.

Beautiful cobalt blue and gold porcelain with an attached candle holder. The two inkwells have hinged lids with pointed finials. A pen tray is across the front. Double loop handles on the sides. England, c. 1865. 5" x 11". $1,400-1,600.

Kidney shaped mirrored tray with two inkwells and a pen holder attached. Pink and white porcelain with brass lids. Marked "France." c. 1910. 4 1/2" x 9", 2 3/4" tall. $200-300.

Elaborately decorated with inkwells on each side that have loose covers. The cherub standing in the hole in back is a seal; the other hole in the back is for a personal seal or a stick of sealing wax. In the middle is a small candle holder and in front are two quill holes. In the center is an oblong compartment for pen nibs, wax wafers, or other accessories. France, c. 1880. 4 1/8" x 7", 2" tall. $1,000-1,400.

Yellow porcelain on a brass latticework base. The matching hinged lid with brass mounts has an applied rose on the top. Floral designs are displayed on a white background and framed in red. A pen rack is across the front. Marked "P.P." France, c. 1900. 3 1/8", x 5 1/2", 4" tall. $200-250.

Celedon green hexagonal inkwell with a matching loose cover. Decorated with pink flowers. c. 1900. 3 1/8" in diameter, 2 3/4" tall. $95-125.

Capo-di-Monte inkwell that has a loose cover with a gold finial. Nude children at play in a rose garden are around the sides. The top has two holes for quills or pens. Marked with a crown and an N on the bottom. Italy, c. 1885. 3" x 4", 2 1/4" tall. $350-500.

Six-piece desk set by Gerand Duffaissein ET Morel, France. Beautifully painted holly on porcelain. Inscribed on the bottom "CFH over GDM. France." (Charles Field Haviland). Mark for 1882-98. The set consists of inkwell with a loose lid, 4 1/4" in diameter; paperweight, 3 1/2" x 5 1/2"; rocker blotter, 2 3/4" x 5 1/4"; pen tray, 4 7/8" x 9 3/8"; and pen, 8" long. $600-800 set.

Irregular shaped base in white porcelain with red and blue applied flowers. A gold trim adds to the decoration. The loose cover on the inkwell has a finial. Marked "Crown PMP 1827. Made in GDR [German Democratic Republic]. Hand painted." 5 1/4" x 6 1/4". $125-195.

Faience inkstand brightly colored with white, green, and dark blue stripes; the design is enhanced with red flowers. The removable lid has a yellow finial and rings of yellow and blue. Marked "Alordin, France." c. 1910. 2 7/8" x 3 3/8", 3" tall. $125-175.

Italian faience with colorful flowers framed in a yellow border. c. 1900. 2 1/4" square, 2 5/8" tall. $75-125.

Triangular shaped blue and white porcelain inkwell with a loose cover. The tray across the front holds writing accessories. Marked "Koninklijke Porceleyne Fles, Delft, Holland. MV 1456 C.D." 6 1/2" x 9". $250-350.

White cornucopia on an oblong base. On the top are two inkwells with loose covers, and two quill or pen holes. On a paper label on the bottom is "Mottahedeh." Italy, c. 1925. 2 1/2" x 7", 4 1/2" tall. $100-150.

Cobalt blue porcelain trimmed in gold. A wreath encircles the funnel dip hole on the top; around the sides are floral and leaf decorations with a scalloped border. France, c. 1900. $100-150.

Colorfully decorated faience. The loose cover has a decorative finial. #38 and #1981 are on the bottom. Marked "France." c. 1890. 3 1/2" x 3 1/2", 4 1/4". $250-275.

Footed inkstand with a seashell flanked by a pair of inkwells. Marked on the bottom "Shofu China. Made In Occupied Japan. Ardalt 6113." White enhanced with green and gold. c. 1947. 3 3/4" x 6 1/2", 3 5/8" tall. $200-300.

White porcelain decorated with green leaves and blue flowers with a gold trim. Removable lid. Three quill holes in the top. England, c. 1920. $100-150.

Square porcelain with a pretty scene of a shoreline and palm trees. The cover is loose. Marked on the bottom "Hand painted Nippon." Japan, c. 1910. 2 3/4" x 2 3/4", 2 1/2" tall. $200-300.

Highly glazed cream-colored porcelain with a green scalloped border. The lid has a hinged brass collar. France, c. 1900. $100-150.

Porcelain with a glaze that gives the appearance of pottery. The single inkwell has a domed hinged lid. The base has a long tray for pens and two depressions in the back for pen nibs or other accessories. Inscribed in gold on the bottom: "Souvenir De France. Mehun s/r Yèvre. From Ralph. October 1918." Signed "A. Monganaste." On the right side marked "Venes." 4 3/8" x 7 3/8", 2 3/8" tall. $450-550.

Brightly colored ceramic with a mushroom shaped lid. Painted leaves and red berries are on the four sides, and a bird with leaves and berries are on the lid. Enhanced with gold flowers scattered on the red background. The brass collar is hinged. c. 1890. 2" x 2", 2 3/4" tall. $200-300.

Glazed red porcelain in the shape of a sofa. The two inkwells have loose covers. In the center is a depression for pen nibs. Inscribed on the bottom "Sarreguemines, France. 4610 531 P Bv." c. 1920. $300-400.

Delicately decorated porcelain with a saucer base. Swags of red roses are tied up with a bow. Enhanced with a finial on the loose cover and rings of gold. On the bottom is a spread eagle and "Josephine." France, c. 1880. 4 3/4" in diameter, 3" tall. $225-300.

A porcelain bowl with a loose cover sits on a saucer base. White background with red flowers, green borders, and a metallic gold trim. Marked on the bottom "Dresden 80h." Germany, c. 1900. 4 1/8" in diameter, 2" tall. $225-300.

Lovely turquoise inkwell attached to a square plate. The inkwell has a quill hole and a loose lid. Decorated with dainty flowers in gold frames. On the bottom is "Decor Main Porcelaines Champs Elysee Paris France. Limoges France." c. 1910. Porcelain. 5 1/4" x 5 1/4", 2 3/4" tall. $200-300.

White porcelain tea kettle style inkwell decorated with orange chrysanthemums and yellow and green leaves. c. 1880. 2 3/8" x 3", 2" tall. $700-1,000.

Oriental tea kettle type with a frog on the top. Decorated with cherry trees and flowers in shades of pink and red on a white background. The top and spout are dark red and enhanced with swirls of gold. c. 1880. 2 3/4" x 4", 3" tall. $700-1,000.

Triangular shaped in a colorful chintz pattern. Marked on the rim of the hinged lid "E.P.N.S.," which stands for electro-plated nickel silver. #1063 is on the bottom. England, c. 1910. 1 1/2" x 1 3/4", 1 5/8" tall. $250-350.

A wheat basket stands on three black legs. The cover has a small brass ring for a finial. The pink inkwell is standing on a blue base with a shell shaped dish on one side and a turquoise jardiniere on the other side. France, c. 1850. $400-600.

Porcelain deep-dish inkstand with three quill holes surrounding the inkwell. Hand painted flowers are framed in a gold diamond-shaped band around the sides. The removable lid has a gold finial. France, c. 1920. 4" in diameter, 2" tall. $150-250.

Porcelain inkwell with scenes around the sides of people on horseback. The lady is riding sidesaddle. Trimmed in gold. Limoges, France, c. 1900. 2" x 2", 2" tall. $85-150.

Bulbous shaped porcelain inkwell with matching hinged lid. Alternate panels of dark green and white decorated with metallic gold leaves and flowers. France, c. 1870. Bronze hinged collar. $200-250.

White porcelain decorated with pink flowers and trimmed with gold. Germany, c. 1890. 2 3/4" x 2 3/4", 3" tall. $100-150.

Unusual hexagonal inkwell. Flowers are around the top and plate with cobalt blue accents. Marked on the bottom with a blue "N" in a shield. Spain, c. 1900. Porcelain. 4 1/2" x 8", 3" tall. $95-150.

White porcelain with a handle in the center. In addition to the two inkwells, there are four quill holes on the top. Decorated with small blue flowers with yellow centers. The gold trimmed base has scroll feet. Germany, c. 1880. 5 1/4" x 9", 5" tall. $300-500.

White porcelain with a matching lid that has a hinged brass collar. Hand painted scene of a lake, mountains, and flowers. England, c. 1900. 4 1/2" x 5", 3 1/2". $125-150.

Porcelain with a square bottom and a round mushroom shaped lid. The brass collar is hinged. Hand painted with a bird among flowers. Dark green enamel covers the top of the inkwell. France, c. 1890. 1 5/8" x 1 5/8", 3" tall. $125-150.

White porcelain inkwell that has a matching lid with a hinged brass collar. Hand painted turtle with flowers and leaves. England, c. 1900. 2 1/4" x 2 1/4", 3" tall. $125-150.

A bunch of grapes are resting on a grape leaf. The lid is removable and opens to reveal a loose inkwell and sander inside. Impressed on the bottom "H H 1108." Germany, possibly Mettlach, c. 1890. Yellow glazed pottery touched with metallic gold. 5 3/4" x 7 1/4", 3" tall. $125-175.

Scallop-edged porcelain trimmed in lavender and gold. The oblong platter shaped base holds an inkwell and a sander. Germany, c. 1890. 6 5/8" x 9 1/4", 2" tall. $125-175.

Inkstand in the form of a chest of drawers. The inkwell on the right side and the pounce pot on the left are removable. There are two quill holes in the middle. England, 1880. 2 3/8" x 4 5/8", 3" tall. $150-250.

Cobalt blue inkwell with a loose lid. A hole in the side holds a quill or pen. A tan design encircles the well and decorates the top of the lid. Marked "Doulton, Lambeth." England, c. 1890. 3 3/4" in diameter, 2 1/2" tall. $350-500.

Royal Doulton inkwell with a reservoir in front. Cobalt blue with light tan decorations around the sides. The swag encircling the inkwell has faces interspersed with scrolls. On the bottom are a crown and a lion with R.D. in a circle and 6599 BB5. England, c. 1880. 3 1/8" x 3 5/8", 3" tall. $800-1,200.

Lusterware in the form of a woman wearing a strapless gown decorated with flowers. The top half is removable. Marked on the bottom "Noritake. Made in Japan." c. 1925. 3 1/4" in diameter, 4 3/4" tall. $125-175.

A pounce pot is on the right and an inkwell with a loose cover on the left. The tray has a tan border with white leaves. Impressed on the bottom is a mark with #43. England, c. 1895. Porcelain. 5 5/8" x 7 5/8", 3 1/4" tall. $225-325.

Green lusterware with four quill holes. The loose cover has a white flower on top. There is a matching lusterware insert. Marked "Germany" on the bottom. c. 1925. 3 3/8" x 3 3/8", 3" tall. $100-150.

Green lusterware with pink and yellow applied flowers. Underneath the removable cover is a matching lusterware insert. There are two quill holes. Marked "Germany." c. 1925. 4 1/2" in diameter, 2 3/4" tall. $125-175.

Brightly painted Italian faience with fruit, berries, and flowers. Winged figures are on the sides. The loose lid has a finial. c. 1880. 7" x 7 1/2", 3 3/8" tall. $450-550.

Raised figures of women and children in a garden setting are around the sides. The top has two quill holes and is decorated with flowers. The loose lid has a gold finial. Hand painted. Marked "G. B. R. Capodimonte 916 Italy." c. 1910. 3" x 4 1/4", 3 1/4" tall. $400-500.

Faience inkwell in the form of a hat that has a hatband with a bow across the front. The loose cover has a picture of a seated boy blowing a horn. Brightly painted in blue and yellow. Marked on the brim of the hat "HR Quimper." Mark used 1895-1922. Quimper is a town in France where several potteries were located. 5 1/2" x 6", 3" tall. $350-500.

Hand painted Capodimonte type porcelain. Various fruits and leaves decorate the red band that encircles the sides. Loose lid with a finial. Marked "1535/131R Italy." c. 1890. 2 7/8" x 4 1/8", 3" tall. $175-250.

Blue and white faience inkwell with a painted thistle on the front and back. The loose lid has a handle. There are four quill holes on the top. Marked on the bottom "E. Gallé Nancy." France, c. 1890. 2 3/4" x 2 3/4", 3 1/2" tall. $1,000-1,500 (rare).

Faience inkstand with a woman dressed in blue wearing a yellow apron. Decorated with a green, red, blue, and yellow floral band. Edged with a scallop and dot design. The two inkwells have removable lids. Marked on the bottom "France HR Quimper." This mark was used 1895-1922. 3 1/2" x 6 1/2", 3 7/8" tall. $900-1,200.

Round porcelain with a nickel silver collar and sliding lid. A dark blue floral and scroll pattern on a creamy white background. America, c. 1880. 4" in diameter, 2" tall. $200-300.

Footed inkwell with a loose lid that has a gold finial. An orange flower on a white background is trimmed in dark blue and gold. Marked on the bottom as well as under the lid "Royal Crown Derby. Made In England. 2451. Bone China." c. 1910. 2 3/4" x 2 3/4", 3 7/8" tall. $600-800.

A brightly painted Staffordshire basket with an inkwell on the left and a sander on the right. A red bow decorates the handle. England, c. 1880. 1 1/2" x 3", 3 3/4" tall. $125-200.

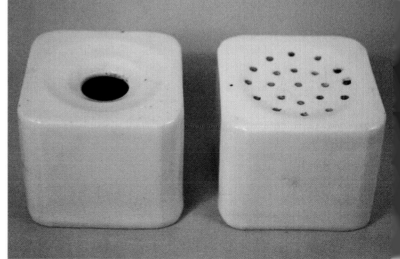

White porcelain cube-shaped inkwell and matching sander. Made by Bing S. Grondal. Denmark, c. 1900. 2 1/2" x 2 1/2", 2" tall. $150-200.

Ribbed milk glass with a brass cover. On the left side is a hinged lid with a finial; on the right is a round quill hole. France, c. 1890. 2 3/4" in diameter, 3" tall. $275-325.

Inscribed on the bottom "Erphila Ink Girls." A channel for a pen is across the front. Germany, c. 1920. 2 3/4" x 5 1/8", 4 1/4" tall. $200-300.

Staffordshire inkstand in the shape of a little girl's vanity table with a mirror. The two inkwells have loose lids and matching inserts. #3259 is on the bottom. England, c. 1900. 2 1/4" x 4 5/8", 5 3/8" tall. $150-250.

Pink heart with a red rose on the loose cover. There are two quill holes on the top. Enhanced with metallic gold trim. Marked on the bottom "France Hand Painted." c. 1900. 3 1/4" x 3 1/4", 2 3/8" tall. $200-300.

Staffordshire pottery bird nest with three speckled eggs. The mother bird is alarmed by the presence of a green snake on the side. England, c. 1870. 3" in diameter, 3" tall. $350-450.

Staffordshire pottery with a mother bird and two babies in a nest. Colorful beaded overlay. England, c. 1860. 2" in diameter, 2 3/4" tall. $200-250.

Blue porcelain with applied white flowers and beads attached to a saucer base. The lid is hinged with a brass collar. Marked "02169 Made in Germany." c. 1910. 4" x 4", 2 3/4" tall. $150-200.

A washerwoman is holding a bar of soap in her right hand. The tub is the inkwell and on the back corner is a round tube for a quill or pen. Germany, c. 1890. 3" x 3 1/4", 4 1/2" tall. $175-250.

George Washington, the first president of the United States, is seated on a white horse tipping his hat. "Washington" is written in the white oval on the front. The figure of Washington and the horse lift off to reveal an inkwell and a sander underneath. America, c. 1880. Colorfully painted porcelain. 2 3/4" x 5", 6 1/2" tall. $500-700.

Open view of the George Washington stand showing the inkwell and the sander.

A drummer wielding a pair of drumsticks is sitting on a cushion between two drums. The drum on the right holds the ink and the one on the left is a sander. The loose sander has a stem on the bottom that fits in a hole on the top of the base. The drum lids are removable and are marked underneath 15 and 23. The composition stands on four scroll feet. Germany, c. 1885. Porcelain. 3" x 5", 3 7/8" tall. $500-750.

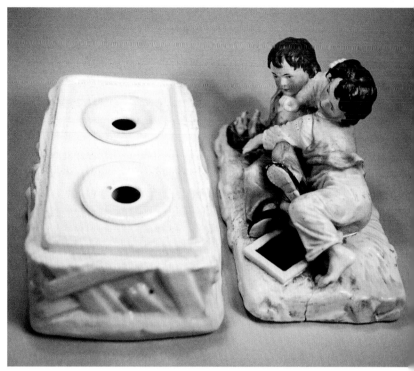

Charming Staffordshire inkstand displaying two boys on the way home from school (note the slate on the right side), fighting over a nest that holds two baby birds. The top of the composition lifts off and exposes two inkwells. England, c. 1890. 4" x 7", 6" tall. $700-900.

Open view of the two boys and the bird's nest.

A romantic scene with a woman seated at a harpsichord while the man beside her has his arm around her shoulder. The top half of the inkstand lifts off to reveal a sander and an inkwell. Germany, c. 1880. Porcelain. 2 1/2" x 4 3/4", 4 3/4" tall. $300-400.

The two inkwells have loose lids with knob finials. A pen rest and a dish to hold pen nibs are across the front. Pink roses are painted on a soft green background. The white leaf and scroll border is brushed with gold. Dresden style. Germany, c. 1900. Porcelain. 6" x 8 1/2". $200-300.

Open view showing the tray under the chess-playing trio. Note the mustachioed man's face on the front of the inkwell and sander.

Beautifully detailed inkstand with the figures of two court ladies and a gentleman playing chess. The whole composition lifts off to expose a tray underneath that holds an inkwell and a sander. Germany, c. 1880. 4 3/4" x 6 3/4". $600-800.

Beautifully crafted inkstand with many intricate details. A family scene is depicted with a mother showing off her baby. This is the cover; the whole composition lifts off to expose a tray underneath that has an inkwell and sander attached. The two inserts are loose. Germany, c. 1880. Porcelain. 6 1/2" x 8 1/2", 8 1/4" tall. $600-800.

Bottom part of the inkstand with three people admiring a baby. Note the mustachioed man's face on the front of the inkwell and sander.

White porcelain with leaves edged in gold. A sander is on one side and an inkwell on the other. The matching loose covers have ring finials. A curled leaf at the top forms a cup that can be used to hold pen nibs. Inscribed "C. Tielsch & Co." Germany, c. 1890. 6 3/4" x 10 1/4", 5 1/2" tall. $400-475.

Hand painted with pink and red flowers enhanced with a gold border. In addition to the inkwell there is a pounce shaker and a pen drainer. All three are missing lids. The handle is in the shape of a swan. England, c. 1840. $700-850.

Lovely inkstand with dainty pink, yellow, and blue flowers enhanced with an elegant gold trim. On each side is an inkwell and in the center a pounce pot. All three have loose inserts. The separate covers have decorative finials. There are four quill holes between the inkwells. Initialed on the bottom with a J with a dot above and below, and a P. Paris porcelain by Jacob Petit. France, c. 1830-1862. 8" x 15". $1,500-1,800.

Porcelain Rococo style with a blond figure sitting in the center holding a letter in his hand. The two inkwells have loose lids with finials. France, c. 1880. 5" x 8 1/2", 6 1/2" tall. $500-800.

White porcelain in the form of seashells. The revolving snail type inkwell is attached to a brass bracket that has a stopper disc in the back and a pen rack on the top. France, c. 1880. 6" x 6 1/2", 4" tall. $400-500.

Delft-like inkstand with a pair of cherubs sitting in the middle. The cherub on the right side is holding a letter in his hand. The inkwells have loose lids with finials. Painted in a pretty blue and white design with a sailboat framed on the front. Europe, c. 1890. 5 3/4" x 7 1/2". $300-500.

Porcelain revolving snail with a saucer base. The inkwell is attached to a brass bracket that has a stopper plate in the rear and a pen rack on top. Decorated with a pink stripe and a dainty floral wreath. Signed on the bottom "Mc." France, c. 1880. 5" in diameter, 3 1/2" tall. $350-450.

Porcelain revolving snail on a saucer base. The inkwell is attached to a brass frame with a round stopper disc in the back and a pen rack on the top. The inkwell swings down to open and is closed by rotating up against the stopper disc. Yellow flowers and green leaves are on the inkwell and the edge of the saucer. France, c. 1880. 3 1/2" in diameter. $300-450.

The revolving milk glass snail inkwell is attached to a brass saucer-shaped base. The supporting bracket has a stopper plate and a pen rack. France, c. 1880. 5 1/2" in diameter, 3" tall. $300-400.

A milk glass snail inkwell that is encircled with a painted floral design. The two brass inkwells on the sides have loose covers. A pen rack is in the rear and there is a recessed tray in the front for pens. France, c. 1860. Black wooden base. 4" x 8 1/2", 4 1/4" tall. $400-600.

Three milk glass revolving snail inkwells. The ends of the wells are painted red, blue, and black, to indicate the color of the ink inside. When closed, the open tops rest against stopper plates; when one is rotated down, all three open. Surmounting the inkwells is a scalloped dish to hold pen nibs or wax wafers. France, c. 1860. The saucer shaped base measures 7 3/4" in diameter. 5" tall. $1,000-1,500.

Double cast iron inkstand fitted with two milk glass snails. The pen rack on the top holds four pens. H. L. Judd Manufacturing Company, Wallingford, Connecticut. America, c. 1880. 3 3/4" x 5 1/2", 4" tall. $400-600.

Three milk glass snails on a lazy susan stand. The cup at the top holds pens nibs or wax wafers. The round base is heavily engraved. Landers, Frary, and Clark, New Britain, Connecticut. America, c. 1880. 6 3/4" in diameter, 5 1/2" tall. $800-1,000.

Two milk glass snails are affixed to an iron frame. They revolve downward to open, and up against the stopper plate to close. The top part of the frame forms a pen rack that holds two pens. Peck, Stow, and Wilcox, Southington, Connecticut. Hull design stopper disc. America, c. 1880. 5" x 7 1/2", 4 1/2" tall. $400-500.

Revolving milk glass snail attached to a copper plated brass frame. In the rear is a pen rack with a stopper disc. France, c. 1890. 5" in diameter, 2 5/8" tall. $300-450.

Unusual Damascene revolving inkstand with a candle holder and a pen rack. The snail inkwell rotates down to open. The tripod stand has a rear leg that is hinged. c. 1880. Iron with a bronze finish. 3 3/4" x 6", 7 3/4" tall. $500-800.

Revolving milk glass snail on a brass stand. At the top of the iron frame is a pen rack. The square base has pointed ends and is covered with an intricate design. Wolcott Hull. America, c. 1880. 4 3/4" x 7", 4" tall. $300-450.

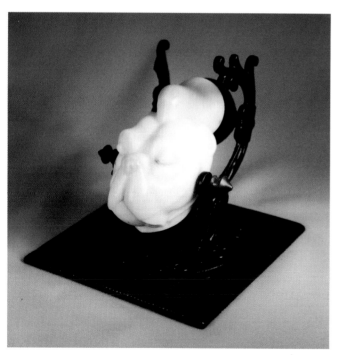

Revolving milk glass inkwell in the shape of a bulldog's head. The iron frame has a pen rack on top and an affixed stopper plate. Landers, Frary and Clark, New Britain, Connecticut. America, c. 1880. 4 5/8" x 4 5/8", 4 1/4" tall. A piece of the iron frame is broken off in this photo. $500-800 (if in good condition).

Two milk glass revolving snails. There is a pen rack at the top of the stand. The iron base is decorated with a scroll design. Landers, Frary and Clark. America, c. 1880. 4 1/4" x 6 1/2", 3" tall. $400-550.

A milk glass snail in an iron frame. The stand has an attached round stopper plate on the back and a pen rest on the top. Wolcott Hull, Meriden, Connecticut. America, c. 1880. Iron. Triangular shaped base with a foot in the back. 4 1/2" x 6", 4 1/4" tall. $300-450.

A milk glass snail is attached to the iron frame. The well pivots down to open and rotates upward against the disc in the back to close. Footed stand with a pen rack on top that holds two pens. Landers, Frary, and Clark, New Britain, Connecticut. America, c. 1880. 5 1/2" x 5 1/2", 5" tall. $300-500.

Inkstand with a single pressed glass snail bottle. The pen rack across the bottom has a pierced leaf decoration. Marked on the back of the stopper disc, also on the bottom of the frame: "Patented November 25, 1879." H. L. Judd Manufacturing Company, Wallingford, Connecticut. America. 3" x 4", 3" tall. $250-350.

Double with clear pressed glass snail type inkwells. Irregular shaped base with etched floral decorations. The pen rack on top holds two pens. Peck, Stow, and Wilcox, Southington, Connecticut. America, c. 1880. Iron. 4 7/8" x 7 3/4", 4 1/4" tall. $400-500.

Double revolving with floral pressed glass snails. An openwork pen rack is across the front. H. L. Judd Manufacturing Company, Wallingford, Connecticut. America, patented November 25, 1879. Iron frame. $300-450.

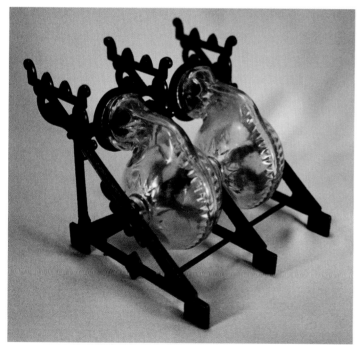

Inkstand with two pressed glass snail-shaped inkwells. The pen rack on the top holds three pens. Peck, Stow, and Wilcox, Southington, Connecticut. America, c. 1880. Iron frame. 4 3/8 x 5", 4 1/8" tall. $300-450.

Iron frame fitted with a single snail inkwell. The rack across the top holds four pens. H. L. Judd Manufacturing Company, Wallingford, Connecticut. America, c. 1880. 2 3/8" x 3 1/2", 4" tall. $300-400.

Cast iron stand fitted with a cobalt blue snail inkwell. At the top is a pen rack that holds two pens. Marked "Pat. May 14, 1878." Landers, Frary, and Clark, New Britain, Connecticut. America. 2 1/4" x 4 1/2", 3 1/2" tall. $350-500.

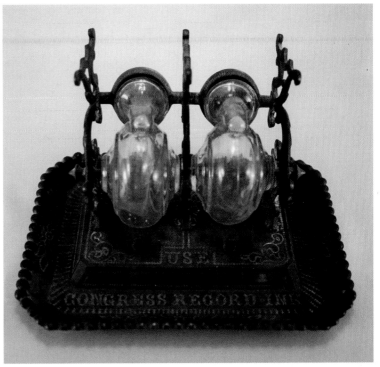

A pair of snail type inkwells in a cast iron frame. The pen rack on top holds three pens. The message inscribed on the base in front is "USE CONGRESS INK." Wolcott Hull. America, c. 1885. 5 1/4" x 8", 4 1/4" tall. $400-550.

Inkstand with two snail type inkwells. The attached pen rack has round stopper discs and holds three pens. The oblong base has rounded corners with scallops around the sides. America, c. 1880. Cast iron with gold paint. 5 1/2" x 8 3/8", 4 1/2". $300-450.

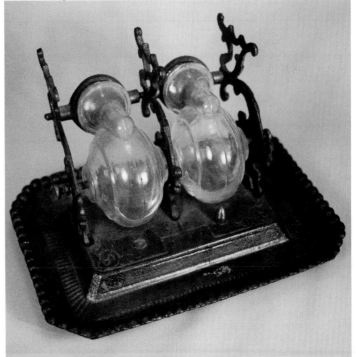

Two snail type inkwells. The pen rack on the top is part of the frame and holds three pens. This is basically the same inkstand as the "Congress Ink," however it does not carry the advertisement. Wolcott Hull. America, c. 1880. 5 1/8" x 7 5/8", 4 1/4" tall. $350-450.

A clear snail type revolving inkwell. The pen rack on top holds three pens. Peck, Stow, and Wilcox, Southington, Connecticut. America, c. 1880. Iron frame. 2 1/2" x 4 1/4", 4 1/8" tall. $300-350.

Ornate stand that holds a clear glass snail inkwell. A pen rack is across the top and another one across the lower front. Wolcott Hull, Meriden, Connecticut. America, c. 1880. Iron frame. 2 3/4" x 4 3/4", 4 1/2" tall. $350-450.

Inkstand with a nautical motif. There is an anchor attached to ropes on each side and a stopper plate in the back in the shape of a ship's wheel. A pen rack is at the top of the stand and across the lower front. Clear snail type bottle. America, c. 1880. Iron. 2 1/2" x 4 1/4", 4 1/4" tall. $300-450.

Ornate iron frame with a clear glass snail. The pen rack on top holds three pens. Marked on the bottom "Tatum's Revolving." Tatum Manufacturing, Cincinnati, Ohio. America, c. 1880. 2 1/4" x 4 7/8", 4 3/8" tall. $350-450.

Ornate frame with a pair of snail inkwells. The pen rack on the top holds three pens. On the bottom of the stand is embossed "Tatum's Revolving." Both wells have "Tatums" pressed in the glass. Tatum Manufacturing, Cincinnati, Ohio. America, c. 1880. Iron frame with a bronze finish. 3 5/8" x 5 1/2", 3 3/4" tall. $350-550.

Pair of glass snails in an iron frame. A pen rack is on the top. Peck, Stow, and Wilcox, Southington, Connecticut. America, c. 1880. 4" x 7 1/4", 4 1/4" tall. $300-500.

Eastlake style (popular in Victorian days) with a hexagonal shaped crystal inkwell that has a cone shaped loose lid with a ball finial. A match container on the right side has a matching lid with a ribbed striking surface underneath. The pen rack holds two pens. America, c. 1880. Iron with a bronze patina. 4 1/8" x 6 1/8", 5 1/4" tall. This is a double collectible. $400-600.

View of the inkstand showing the match compartment.

Inkstand with a matchbox as part of the design. The pressed glass inkwell has a loose cover with a finial. The match container is on the right side and has a loose lid with a matching finial. When the lid is turned over, there is a ribbed surface underneath for striking a match. The pen rack in the back holds three pens. America, c. 1880. Iron with a bronze patina. 3 7/8" x 7 1/2", 3 7/8" tall. $450-600.

Inkstand with a single octagonal crystal inkwell. The lid is attached to the frame and when the oval plate in front is pressed down, the lid is lifted. The pen rack holds three pens. On the base is an Oriental garden scene. A woman's face is embossed on the top of the oval lever bar and the top of the lid. Marked in the back, "Brooks & Co. Rogers Ford, Pa" and on the bottom, "Pat. Jan. 30, 1883." America. Iron. 5 5/8" x 6 1/2", 4 1/2" tall. $450-600.

Above: A pierced railing is around the sides and back. The single well has a hinged lid. A book, horn, and flowers are the decorations. Unusual shape, square angled corners and rounded front. c. 1900. Brass with an antique finish. $300-400.

Left: Pressed glass inkwell in an iron stand. The lid is attached to a bar on the frame. On top is a pen rack that holds three pens; in the rear is a slot to hold a blotter, postcard, or letters. The backplate has a design of open arches. The base is rounded on the corners and pointed on the sides. Marked "Pat Nov. 25, 1879." America. Iron with a bronze patina. 4 1/4" x 5 5/8". $250-350.

Double inkstand with a pen rack on the top for three pens. A slot in the rear holds a blotter, postcard, or letters. The two pressed glass wells have lids that are attached to a bar on the frame. America, c. 1890. Iron with a bronze patina. 4 1/2" x 7", 5" tall. $300-400.

Roll top stand with a pen rack on top. The two individual doors roll up and back to expose a pair of square crystal inkwells. A wide space in front holds a letter opener or pens. There is an Oriental scene on both covers; the figure of a woman is on the left cover and a man is on the right. America, c. 1890. Iron with a bronze patina. $300-400.

Single inkstand fitted with a round paneled glass inkwell. The frame has a coiled fish design on the sides topped by spires. The lid is attached to the frame. In front is a compartment with a hinged lid for pen nibs or stamps. Number 5 is on the bottom. Iron with a copper finish. America, c. 1880. 3 3/4" x 5", 5 1/2" tall. $200-275.

Brass set with glass jewels. The dome shaped inkwell stands on a pedestal and has a double spiked finial on the hinged lid. There are six clear jewels around the top and twelve green jewels around the base. 6 1/2" in diameter, 5 5/8" tall. c. 1910. $300-450.

Two swirl glass inkwells that have separate covers. In the center is a hinged pen nib or stamp box. A floral design is on the lids and the ends of the stand. A channel for pens is in the back as well as the front. Mark with "Bradley & Hubbard Mfg. Co. B. & D. 6277." America, c. 1900. Brass. 4 7/8" x 10 1/2". $300-400.

Square wooden base overlaid with a copper frame. Fitted with a triangular shaped clear glass inkwell. The cover is attached to the pen rack in the back. Stands on four paw feet. Marked "Vac & C." on the bottle. America, c. 1910. 5 3/8" x 6", 6" tall. $150-200.

A pair of swirl glass inkwells are seated in the base. The lids are attached to the frame. The pen rack holds three pens and is separated with two openwork disc with a T in a diamond. Tatum Manufacturing, Cincinnati, Ohio. America, c. 1880. Cast iron. 3 7/8" x 6 1/2", 4" tall. $350-450.

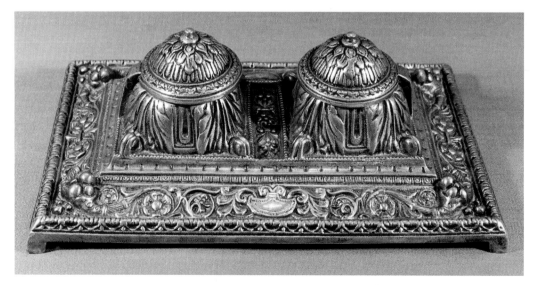

Nickel-plated over brass. The lids are hinged on the outside edge and open sideways. Cornucopias decorate the corners and there is a leaf and scroll design covering the entire subject. c. 1900. 6" x 9 5/8". $300-450.

This inkstand has an unusual pen rack at the top: a coiled spring set in an iron base. The footed stand has a swirl glass inkwell with a lid that is attached to the frame and slides to the side to open. America, c. 1880. Cast Iron. 6 1/4" x 6 1/4", 4" tall. $300-400.

oll top by "Bradley and Hubbard Mfg. Co. #6076 Pat' Apl'd For." Two milk glass serts. Openwork pen tray across the front. Meriden, Connecticut. America, 1885. Iron with a bronze patina. $250-400.

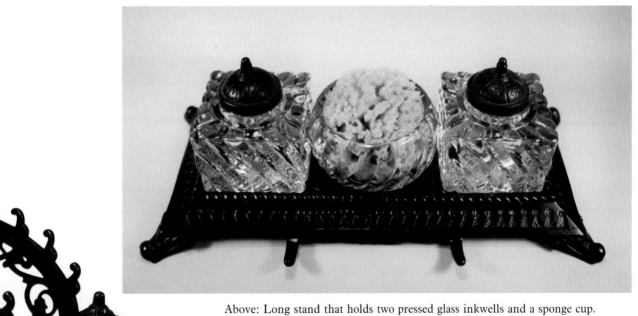

Above: Long stand that holds two pressed glass inkwells and a sponge cup. The sponge was used to moisten the glue on envelopes and stamps. The lids are removable. There is a pen rest across the front. America, c. 1890. Cast iron. 5 1/2" x 11". $300-400.

Left: The ornate pen rack holds three pens. The pressed glass inkwell is inscribed on the bottom "Pat Dec. 11, 77." The lid is removable. America. Iron. 4 1/2" x 4 1/2", 5 1/4" tall. $300-400.

Cast iron frame with a pen rest across the front that holds three pens. The pressed glass inkwell has a loose cover. The irregular shaped base is decorated with a saw tooth design. Inscribed on the base "B. D. Pat. Sept. 1, 1875." America. 6" x 7". $200-300.

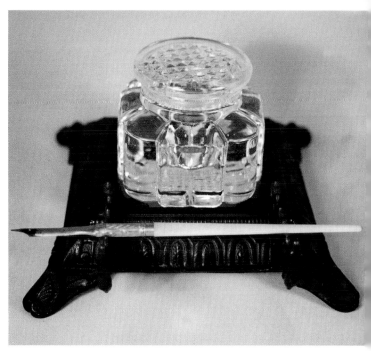

Footed inkstand with a pen rest across the front that holds four pens. The pressed glass inkwell has a loose cover. Marked on the bottom of the frame "Patented July 14, 1878." America. 6" x 6". $200-250.

Small desk set consisting of an inkwell, rocker blotter, and a letter stand. Marked "J. B. 10." Jennings Brothers Manufacturing Company, Bridgeport, Connecticut. America, c. 1928. White metal with a bronze coating. Inkwell: 3 1/2" x 3 5/8", 3 1/2" tall. Letter rack: 1 1/2" x 4", 4" tall. Blotter: 1 1/2" x 4", 4" tall. $150-250 for three-piece set.

Two pressed glass inkwells with loose lids fit into holes in the top of the stand. In the middle is a stamp box with a hinged lid, and in front a long channel for a pen. An oak leaf is impressed on each side and on top of the stamp box. Marked on the bottom in a triangle "Bradley & Hubbard Mfg. Co." Meriden, Connecticut. America, c. 1910. Gilded iron. $300-400.

Inverted blossom with a hinged lid. A pen rack is across the front. Decorated with leaves in heart shaped frames. Inscribed on one leg "Pat. 1872." America. Iron. 4 1/2" wide, 21/2" tall. $195-250.

Cast iron with a pressed glass inkwell that has a loose cover. The openwork pen rack in the back is unattached and connects to the base with two hooks. A pen channel is across the front. The decorations consist of a dog running across the front of the base, and grapes and leaves on the pen rack. America, c. 1880. 4 3/4" x 5 1/4", 6" tall. $250-350.

Dome shaped inkwell with a hinged lid. The pen rack on the front holds two pens. America, c. 1880. 3 3/4" in diameter, 2 1/2" tall. $195-250.

Small inkwell with a hinged lid is attached to a base that slants up the back. Panels around the inkwell alternate between a solid and openwork design. Marked "Pat Apl'd For." America, c. 1880. Iron with a bronze finish. 4" x 4 1/2", 2 1/2" tall. $125-175.

Eastlake style six-sided inkwell with a hinged lid. There is a stepped pen rack across the front that holds three pens. Marked on the top behind the inkwell "PAT. SEP. 9, 1878 B.B." America. Cast iron. 4 1/8" x 4 5/8", 2" tall. $195-250.

Thermometer and inkwell combination. The footed stand has a crystal inkwell with a loose cover. The front ornament is the figure of a woman with wings. Marked "Patented 1870." America. Cast iron. 5 7/8" x 6", 8 1/2" tall. $400-600.

Hudson's barometric inkwell with a hinged pewter lid on the font. The base is weighted brass and the pen rack is cast iron. America, c. 1865. 5" in diameter, 5 5/8" tall. $300-450.

A pressed glass inkwell with a loose lid is fitted in an iron base with hoof shaped feet. The stepped pen rack across the front holds two pens. America, c. 1885. Iron. 3 7/8" x 4 1/4". $175-250.

Double stand with a pen rack in the rear that holds three pens. The two pressed glass inkwells have lids that are attached to the frame. Decorated in a floral pattern. America, c. 1885. Iron. 5" x 8 1/4", 4 1/2" tall. $300-350.

Inkstand with two pressed glass inkwells that have loose covers. The pen rack in back is decorated with the figure of a stork in flight. Stands on four hoofed feet. Impressed on the bottom of the bottles, "P.S. & W. Co. Pat. Dec. 11, 77." Peck, Stowe, and Wilcox. Southington, Connecticut. America, Dec. 11, 1877. Iron. 3 3/4" x 6 1/2", 5" tall. $300-450.

This inkstand was advertised in the 1897 Sears, Roebuck & Co. Catalogue for 45¢! A letter or postcard holder is in the back. The pen rack holds three pens. The pressed glass inkwell has a lid that is attached to the frame. America. Iron. 4 3/4" x 4 3/4", 5 3/8" tall. $200-350.

An ornate cast iron double stand. The openwork frame has a postcard or letter rack in the back. The top of the stand is notched to hold three pens. The two pressed glass inkwells have covers that are attached to the frame. America, c. 1895. 4 3/4" x 8 1/2", 5 1/2" tall. $300-450.

Footed inkstand with a base that has pointed ends. Fitted with a crystal inkwell with a cover that is attached to the frame. The pen rack on top holds two pens. A star decorates the round plate in the back as well as the cover. America, c. 1885. Iron. 5" wide. $150-250.

Stand decorated in an incised scroll design. There is a postcard or letter rack in the back and a rack on top that holds three pens. The cover on the pressed glass inkwell is attached to the frame. America, c. 1885. Nickel-plated cast iron. 4 1/2" x 5 1/4", 5" tall. $200-300.

Eastlake style stand with an ornate pen rack that holds three pens. The base is fitted with an octagonal glass inkwell that has a funnel shaped dip hole. Marked on the bottom of the inkwell, "Safety Inkstand. Pat. Apl. 3, 66." America, 1866. Cast iron. 4 1/4" in diameter. $300-450.

Footed base with a scalloped skirt. Two clear crystal inkwells with loose covers are fitted in a recess on the platform. America, c. 1890. Cast Iron. $200-300.

The lids on the two pressed inkwells are attached to the frame. A letter, or blotter, rack is in the back and a pen tray across the front. A floral design covers the subject. Marked on the bottom, "BRADLEY & HUBBARD MFG. CO. PAT' APL'D FOR," and #7013. America, c. 1900. Iron. 6" x 9". $300-400.

Footed stand with a roll top lid surmounted by a pen rack. A knob on the front lifts the lid and exposes four glass inserts: two small funnel shaped inserts fitted inside two regular inserts. A wheel type decoration is on the sides, fans and scrolls on the front, and an engraved design on the back. America, c. 1885. Bronze. 4" x 4 3/4", 4 7/8" tall. $300-500.

Inkstand with two inkwells that have hinged lids. The four outside edges are concave to hold pens. A scroll pattern covers the inkwells and the outside edge of the base. England, c. 1880. 5 3/8" x 9 1/4", 5 3/4" tall. $300-350.

Art Deco stand with a hinged lid. Inscribed on one foot "J. B. 1083." Jennings Brothers. America, c. 1910. Bronze coated white metal. 3 1/2" x 3 1/2", 2 1/2" tall. $125-195.

Leaves decorate the center and the top of the two inkwells. The large depression in the front is for pens. On the back is "351" America, c. 1920. White metal with a bronze finish. 6 1/4" x 11 3/8", 2 5/8" tall. $150-225.

This small footed inkwell displays the graceful lines of the Art Nouveau period. The hinged lid is adorned with a flower. Marked "J. B. 117." Jennings Brothers. America, c. 1900. Bronze finish on white metal. 2 3/4" x 3 1/2", 3" tall. $175-225.

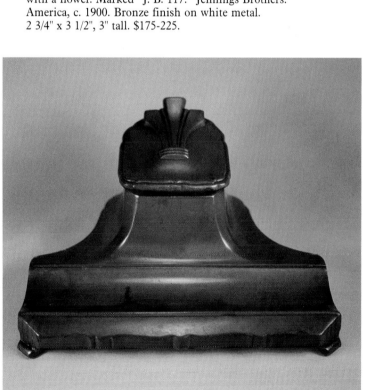

A stack of writing paper, envelopes, and a quill form the base of this inkstand. The inkwell has a hinged lid with a green glass insert. Marked on the bottom "K. & O. Co." (K. & O. Novelties Company, Brooklyn, New York). America, c. 1925. Bronze finish over white metal. 3 3/4" x 6 1/4", 4" tall. $225-325.

Art Deco with a hinged lid. A channel across the front holds pens. Marked "J. B. 3274." Jennings Brothers, Bridgeport, Connecticut. America, c. 1925. Bronze with a verde patina. 4 3/4" x 6", 3" tall. $175-225.

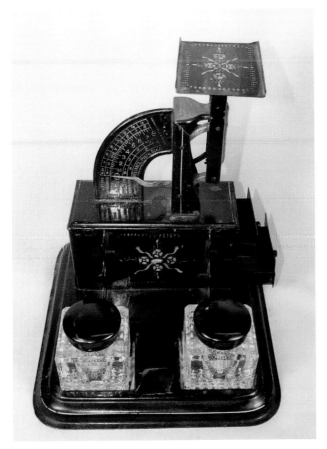

This inkstand has several functions. It is a postal scale, an inkstand with two glass wells with loose covers, and a pen rack. On the right side are two drawers for pen nibs, stamps, and other accessories. Inscribed "Reliance Postal Scale. Patent Pending. Triner Scale & Mfg. Co. Chicago, U.S.A. And Abroad. Pat. Feb. 23, 1904." Tin painted black with white floral decorations. 6 1/2" x 7", 7" tall. $400-600.

Inkwell with a faceted lid and a hinged brass collar. The base has a channel across the front to hold a pen. On the left side is a building with a tall tower that is marked on the lower front "FIRENZE" (Florence). Italy, c. 1920. Alabaster. 4" x 6 3/4", 6" tall. $150-250.

Round pewter counting house style inkstand with a hinged lid and five quill holes around the saucer shaped top. This style has been made since the sixteenth century and was commonly used in early America. Pewter. $200-300.

Capstan style with a hinged lid and five quill holes. America. Pewter. 7 1/4" in diameter, 2" tall. $250-350.

Rococo style inkstand with an openwork back. A crystal inkwell with a loose cover is seated in a frame. France, c. 1890. The well is 1 1/2" square. The brass stand is 3" x 3", 3 1/2" tall. $200-300.

Octagonal footed inkstand with a perpetual calendar. The knobs in front control the date settings and form a pen rest. The inkwell has a hinged lid with a finial. England, c. 1870. Brass. 8" x 8", 4" tall. $400-600.

Two ornate inkwells on pedestals flank a tall perpetual calendar. The lids are hinged. Decorated with lion heads on the sides and the high relief figure of a man's bearded face on the front. Made in France (for the Spanish market), c. 1860. 5 1/2" x 9 1/2", 7 1/2" tall. $800-1,000.

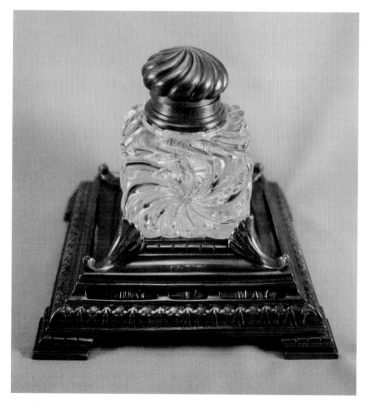

A perpetual calendar is located in the base across the front of the inkstand. The unusual swirl pattern inkwell stands on four scroll feet and has a hinged lid. Signed on the bottom "Towsend & Co. 2870." England, c. 1870. Bronze. Well is 3" x 3". Base is 6 3/4" x 6 3/4", 4 1/2" tall. $500-700.

Pump inkwell on a black wooden base with a tray across the front. The pen rack in the back holds three pens. Porcelain with brass mounts. France, c. 1835. 5 1/4" x 9 1/2". $400-600.

Wooden base with a central pump inkwell flanked by two regular inkwells. All have loose brass covers. In the rear is a pen rack that holds three pens. Marked on the top "Medailde D Argent, 1839." France. Porcelain with brass mounts. 4 1/8" x 7 5/8", 4" tall. $800-1,100.

Wooden base with a pen rack that holds three pens. The central inkwell has a loose cover with a pineapple finial. On the left side is a round box for pen nibs or wax wafers, and on the right is a sander with a pierced cover. Red flowers encircle the porcelain wells. France, c. 1885. Brass lids and pen rack. 4 3/4" x 9 1/2", 4 3/4" tall. $700-900.

A white porcelain pump inkwell. The loose brass cover has an attached porcelain weight that is lowered into the inkwell to control the flow of ink into the reservoir on the lower front. The saucer shaped base has two tubes for quills. Embossed on the top of the cover is "Encrier A. Pompe." France, c. 1845. 5 1/2" in diameter, 4 1/2" tall. $300-400.

A porcelain pump inkwell is in the center, on the right side is a sander, and on the left is an inkwell. The pen rack holds four pens. The channel across the front may hold a penknife for sharpening a quill, a seal, or other accessories. Marked "A te Bocque Brevete." France, c. 1840. Marble base with brass mounts. 4 5/8" x 7 5/8", 5" tall. $1,000-1,500.

Exceptionably nice pump inkwell with a pen rack in the back. The shell shaped tray has serpents coiled around the two quill holes on the sides. France, c. 1840. Gilded bronze with a porcelain inkwell. 5 1/2" x 7", 7 1/4" tall. $2,000-2,500.

Pump inkstand with a brass top. A recessed tray in front is for accessories. There are two quill holes. The lid on the small reservoir is often missing on this type stand. Marked "Medalle D Argent 1839." France, c. 1865. 4" x 5 1/2", 5 1/4" tall. $750-950.

Glass inkwell in the shape of a cloverleaf. A Wedgwood type disc, blue background with white figures, is set in a brass frame on the top. The lid is hinged. There are grooves in the glass on either side to hold a pen. England, c. 1900. 3 1/2" x 4", 2 1/2" tall. $200-350.

Blue and white Wedgewood inkstand with a Wedgewood disc encased in the top of the hinged lid. The brass base and lid are enhanced with a gadroon border. The English registry mark on the bottom tells us this inkwell was made in 1864. Signed "G. Betjemann & Sons." 6 1/2" in diameter, 3 1/4" tall. $900-1,200.

White porcelain inkstand with a clear glass ball on the top that screws into the brass collared well. The font has a brass lid that is attached to the collar by a chain. There are two quill holes. Marked "Encrier Reservois P.B. S.G.D.G. Paris." France, c. 1850. 4 3/4" in diameter, 4 1/2" tall. $500-700.

Pump inkwell. A porcelain weight connected to the lid controls the flow of ink into the small reservoir on the lower front. The brass lift off lid has an elaborate finial. A quill hole is on each side. France, c. 1865. Porcelain. 4" x 5 1/2", 5 1/4" tall. $700-900.

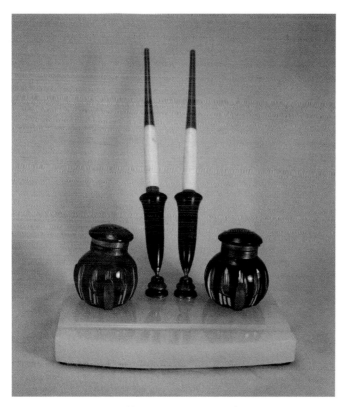

Inkstand with two Bohemian glass inkwells: one cut red to clear, one cut blue to clear. Brass hinged lids. The two pen holders in the middle are fitted with colorful pens. Marble base. 4" x 6". $500-700.

Hexagonal shaped French faience inkwell that stands on three cobalt blue legs. Unusual hinged lid—it is attached to the insert to form one unit. c. 1885. 5 5/8" in diameter, 6" tall. $400-600.

Inkwell with a castellated top is attached to an openwork base. The lid is hinged. Inscribed on the well "W. T. & S." England, c. 1900. Brass. 4 1/2" in diameter. $125-200.

Inkwell composed of a posy of several different kinds of flowers. The daisy at the top is hinged. Unmarked but probably America, c. 1915. Brass with a copper finish. 4 1/2" x 5", 3" tall. $125-195.

Paperweight inkwell. Heavy crystal base with a mushroom shaped lid that is covered with a sterling silver overlay. The hallmark on the rim indicates this was made by Alvin Manufacturing Company. Marked "999/1000 fine. 2125 Patented." America, c. 1890. 3" in diameter, 3 1/4" tall. $350-500.

Ball shaped crystal paperweight inkwell covered with a sterling silver overlay. Beautifully executed in an intricate design of flowers and leaves on a lattice work background. Inscribed on the bottom rim "Black, Starr & Frost. Fine 999/1000 Silver. 762." America, c. 1885. 2 1/2" in diameter, 3" tall. $800-1,200.

Sterling silver overlay on an emerald green glass inkwell. The domed lid is removable. Inscribed on the bottom "General Supply Co. Daniel Son. Jacobus, Ct. Pat. No. 879470." Patent date for 1908. 3" x 3", 2 1/4" tall. $900-1,200.

This inkstand may have been a presentation piece—a prize for winning a bicycle race. Note the bicyclist holding up a victory wreath. The tree stump inkwell has a hinged lid. The bicycle has white tires, a chain drive, and a bell on the handlebars. The composition stands on four scroll feet. Engraved on the bottom "K.M.P.E." c. 1900. Silverplate. 5 1/2" x 7 1/2", 5 3/8" tall. $750-950.

Inkstand with the figure of a golfer taking a swing at a gold ball. The tree stump inkwell has a hinged lid. c. 1900. Silverplated. Base: 3 1/8" x 8". $900-1,200.

An ornate sterling silver overlay of scrolls and flowers cover the top half of the cut crystal inkwell. The lid is hinged. Hallmark for Birmingham, England, 1902. 3 1/8" in diameter, 2" tall. $500-700.

Partners inkstand to be used from either side of a desk. There are pen rests on both sides. The two green porcelain inkwells are decorated with gold flowers and have hinged lids with finials. The round box in the center has a hinged cover and is for wafer seals or pen nibs. The handled oblong base stands on ball feet. Marked "E.P.N.S. Pindar Bros. Sheffield." Electro-plated nickel silver. England, c. 1900. 6" x 10 1/2". $500-700.

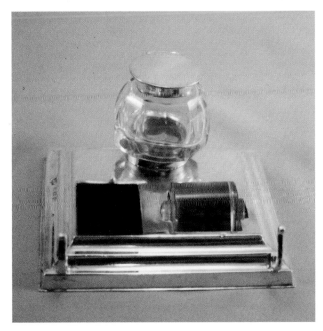

Sterling silver with a clear crystal inkwell. On the left side is a stamp box; a reflection makes it appear black. The lids are hinged. On the right is a clear glass roller for moistening stamps. Across the front is a channel with two posts to hold a pen. Hallmark for Birmingham, England, 1906. 3 1/2" x 4 1/4", 2 1/2" tall. $450-650.

Open view of a silverplated inkstand. The hinged lid has a finial. Across the front is a pen channel and below a drawer for writing paraphernalia. A lion holding a ring in his mouth is on the sides. Stands on four bun feet. Marked "Made In England." c. 1900. This is a copy of an 1800 casket. $300-400.

Sterling silver Art Nouveau inkwell in the shape of a sugar bowl with handles and a hinged lid. Inscribed on the bottom "Shreve & Co. San Francisco," c. 1890. 6" in diameter, 3" tall. $500-800.

Irish pewter. The lid on the inkwell is hinged and on the top are four quill holes. The two drawers with knob pulls hold writing accessories c. 1750. $800-1,000.

Odd shaped, beautifully etched crystal inkwell seated in a sterling silver base. The hinged mushroom-shaped lid is embossed with roses and leaves. Notched in the front to hold a pen. Austria, c. 1900. 3 1/2" x 7 1/4", 3 3/8" tall. $900-1,200.

Art Nouveau inkstand with daisies, buds, and leaves cascading down the side. The large daisy is the inkwell; the lid is hinged and there is an insert. A butterfly rests on the top. c. 1900. Silverplated. 7" x 7 1/2". $200-300.

Ornate footed inkstand with two inkwells. The hinged lids have tulip buds for finials. The back of the stand has an openwork frame with three children at play in the center. On each side are figures of men with their lower halves covered with floral bouquets. Probably Dutch, c. 1880. Continental silver. 5" x 6 3/4", 2" tall. $700-900.

Sterling silver inkwell that has a hinged lid with an enameled floral wreath encircling the top. Hallmark for Birmingham, England, 1912. 3 3/8" in diameter, 2" tall. $350-450.

The figures of a woman and two girls are flanked by cut crystal inkwells. The hinged lids have a floral design on the top. On the sides are openwork leaf decorations. The footed base is decorated with flowers. The long channel holds pens. Europe, c. 1900. Silverplated. 5" x 8". $400-600.

Rococo style silverplated tray with two pressed glass wells seated in frames. The hinged lids extend over the shoulders. Cherubs are displayed among flowers and scrolls. America, c. 1885. $300-400.

Cut crystal inkwell with a hinged lid is seated in a silver frame. The moose head on the front forms a pen rack. Engraved with a bird, bamboo, and leaves. Marked W. & H. (Walker and Hall) in a flag and "Baker Hall, Sheffield, England. 1555." c. 1890. 6" square, 7" tall. $300-500.

Bell shaped sterling silver with a design of grapes, leaves, and vines. Decorative hinged lid with a finial. Marked "3623 S." Hallmark for Gorham Manufacturing Company. c. 1890. 5 1/4" in diameter, 3 1/2" tall. $450-650.

Round sterling silver inkwell with a hinged lid. Marked on the bottom "Tiffany & Co. z280m/0 A." c. 1895. 2 3/8" in diameter. $400-600.

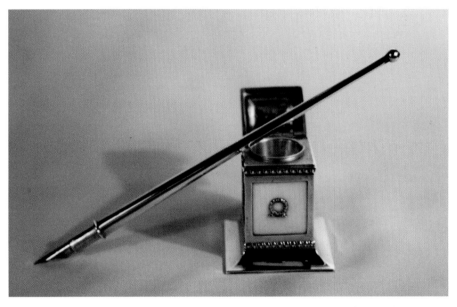

Empire style sterling silver and carrara marble with a wreath on the front. The pen stands in a hole in the center of the lid when not in use. France, c. 1890. 2" x 2", 3" tall. The pen is 6" long. $500-600.

Closed view of the French inkwell showing the pen in place.

Two faceted crystal inkwells with hinged lids are seated in a sterling silver base. The channel across the front is for a pen. Hallmark for Birmingham, England, 1898. 3 3/4" x 5 1/2". $600-900.

Footed inkstand with a face above the pen rest in the back. A brilliant blue crystal inkwell is on one side and a sander with a pierced lid is on the other. Nils Larson Meime. Sweden, c. 1920. Silverplated. 4 3/4" x 6 1/2", 3 1/2" tall. $500-600.

Three seashells are intertwined with seaweed and leaves. The crystal inkwell is encased in a sterling silver bezel. Behind the well is a small candle stand in the form of a flower. The shell shaped tray in the front holds a pen. England, c. 1890. Silverplated, sterling silver mounts. 8" x 10 1/2". $350-450.

Sheffield silverplated gallery tray with four claw feet. Fitted in a frame on the top is a cobalt blue inkwell with a dipping hole in the center and a sander. England, c. 1875. 4" x 7 1/4", 3 1/4" tall. $500-800.

Inkwell with a hinged lid is decorated with a dot and dash pattern around the edges. Marked E.P.N.S., which stands for electro-plated nickel silver. England, c. 1920. 3 1/2" in diameter, 2" tall. $95-125.

Footed platform with the full figure of a standing stag. The crystal inkwells have hinged lids with finials. A pen rack is across the front. Beautifully engraved design with an inscription that reads "Presented To Miss Bisset, United Free Church, Innellan 1904." England. Silverplated. 5" x 7 1/4", 5 1/4" tall. $400-600.

A band of engraved irises are around the sides of this sterling silver inkwell. The separate lid screws onto the base. America, c. 1900. 2 1/4" in diameter, 2" tall. $150-195.

A square crystal inkwell with a hinged lid is seated in a recess on the top of an oddly shaped tray. The rounded corners have a pierced border and are decorated with embossed flowers. Inscribed on the bottom "E.P.N.S. 3005." England, c. 1900. Electro-plated nickel silver. 4 7/8" x 4 7/8", 2 1/8" tall. $150-200.

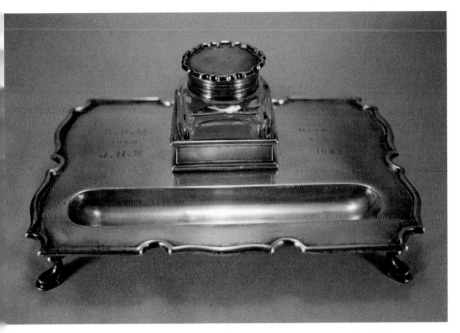

Sterling silver footed inkstand with an applied border. The crystal well has a hinged lid and is seated in a square frame. A pen channel is across the front. Engraved on the top "October Sixth, 1897. E.S.M. From J.H.S." Hallmarks on the bottom inform us that this piece was made in London in 1897. Marked "Goldsmiths Company. 112 Regent street." 5" x 7 5/8", 3" tall. $600-800.

Footed Art Nouveau inkwell with a pair of kissing cherubs on the front. The hinged lid has a flower on top. Inscribed on the bottom "J. B. 518." Jennings Brothers Mfg. Co. Bridgeport, Connecticut. America, c. 1900. Silverplated. 2 3/8" x 4", 2 3/4" tall. $200-300.

Silverplate and brass in the shape of an elephant's foot. The two lids are hinged on the outside edge. Probably England, c. 1890. 4 5/8" in diameter, 3 3/8" tall. $400-500.

Square cut crystal inkwell with a beautiful silver cover in a repoussé pattern. America, c. 1910. 2 3/4" x 2 3/4", 2 1/2" tall. $150-250.

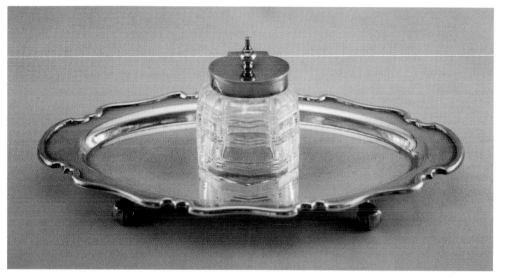

Paneled cut crystal inkwell with a hinged lid sits on a footed oval tray. Several marks are on the bottom, a hand with a fleur-de-lis, HA EA FA 547411, E.P.N., and a clover with #1. England, c. 1909. Electro-plated nickel. 4 1/2" x 6 5/8". $200-300.

Square inkwell with a hinged mushroom-shaped lid. Decorated completely in a pattern of swirls, scrolls, and flowers. France, c. 1890. Silverplate. 1 3/4" x 1 3/4", 2 1/2" tall. $250-350.

Art Deco stand with a mask at the top. The two inkwells have hinged lids. Marked "Austria." c. 1900. 6 3/4" x 9". $150-225.

Footed inkstand with a hinged lid with a finial. A beaded trim is the only ornamentation. Inside are two inkwells and between them a box with a hinged lid for wax wafers, stamps, or pen nibs. England, c. 1880. Silverplated. $300-400.

Sheffield silverplated stand. In the center is a dish on a pedestal to hold wax wafers or pen nibs. On one side is an inkwell on the other a sander. A pen rack is in the back. Stands on four foliated feet. England, c. 1860. 5" x 8". $500-800

A showy stand set with turquoise and ruby colored stones. The crystal inkwell has a hinged brass lid with a filigree design extending over the shoulders. Surmounting the inkwell is a letter holder. Four jeweled feet support the composition. France, c. 1875. Brass. 4 1/4" x 5 1/4", 7 1/8" tall. $500-600.

Lovely Doré bronze inkstand with a pair of green glass inkwells seated in round frames; the decorative lids are removable. In the center is a candle holder and in the back a pen rack. The scalloped-edged tray rests on four pad feet. France, c. 1875. 6" x 8 3/4" (including the handle), 3 1/2" tall. $800-1,000.

Six malachite cabochon jewels are set in bezels: five on the saucer and one large one on top of the hinged lid. France, c. 1850. Gilded brass, the base appears to be hard leather. 5 1/2" in diameter, 2 3/4" tall. $600-800.

Variegated agate inkwell. The lid is hinged with a brass collar. Germany, c. 1910. 2" x 2", 2 1/4" tall. $300-400.

Tiger-eye inkwell with a matching hinged lid. Tiger-eye is a yellow and brown stone (silified crocidolite) that is often used in making jewelry and decorative objects. Germany, c. 1910. 1 1/2" x 1 1/2", 2" tall. $200-300.

Caster style standish with three cylindrical containers: two for ink and one to hold three quills. A handle is attached to the base. Removable lids. Portugal, c. 1820. Brass. 5" x 5", 5" tall. $500-700.

A mythical sea creature is supporting a seashell with his head and tail. The hinged top of the shell opens to expose an inkwell, a sander, and an open compartment in the shape of a half moon. The underside of the lid is shaped like a shell. The composition rests on a marble and bronze base. France, c. 1835. 3 1/2" x 3 5/8, 5 1/2" tall. $800-1,200.

The motif of this stand is chains. A large chain encircles the bottom, a chain around the hinged lid has a malachite gemstone set in the top, and a small chain decorates the front. England, c. 1860. Brass with copper. 5 1/2" in diameter, 3" tall. $500-700.

A brass design covers a slag glass inkwell. The hinged lid opens to reveal a large glass insert. This is just one piece of a matching desk set. Some Tiffany sets consist of twenty or more pieces. Marked "Tiffany Studios, New York. 844." c. 1900. 4 1/8" x 4 1/8", 3 1/2" tall. $800-1,000.

Lily pads with a pair of snakes coiled upward to form a pen rack. A mother-of-pearl dish for pen nibs adds to the decoration. The crystal inkwell in back has a loose cover. France, c. 1890. Brass. 7 1/2" x 9", 4 5/8" tall. $650-850.

Pietra Dura Italian mosaic inkstand. The black onyx base has lilies of the valley, turquoise, and coral flowers on the top. The crystal inkwells are set in frames at an angle on the sides. The brass hinged lids are enhanced with inlaid flowers on an onyx background. The brass frame stands on looped feet. A pen rack is in the back. c. 1865. 4 1/4" x 8 1/2", 2 1/4" tall. $1,000-1,500.

Art Nouveau stand that displays a butterfly on the pen rack in the back. The swirl pattern inkwell has a matching lid with a finial. The hinged collar extends over the top of the inkwell. America, c. 1900. Rolled brass. The tray measures 2 5/8" x 5 1/2". $200-250.

Beautiful urn-shaped inkwell on a pedestal. The oblong tray has handles on the sides. Elaborately decorated. France, c. 1890. Gilt brass. 4 7/8" x 8 3/4", 5" tall. $500-700.

A croquet mallet and a tennis racket are in the back. The hand painted crystal inkwell is seated inside a triangular base. The hinged lid displays the head of a stag on the top. The sides of the base have openwork branches with flowers. England, c. 1910. Gilded brass. 2 1/4" x 4 3/4", 1 3/4" tall. $200-300.

A footed French bronze stand. The inkwell has a hinged lid with a pineapple finial. The lid and outside edges are embellished with scrolls and leaves. The front foot is in the shape of a ram's head. c. 1900. 8" x 9", 4" tall. $400-500.

André Charles Boulle (1642-1732) was one of France's leading architects and cabinet makers who developed the fine art of marquetry. Boulle work was very popular and was imitated during the eighteenth and nineteenth centuries.

Boulle work desk box. Intricate marquetry of tortoise shell and brass. The clear crystal inkwell has a hinged lid and is flanked by two compartments with loose covers that hold pen nibs, wafer seals, or other desk accessories. The drawer in front holds stationery. England, c. 1870. 7 1/2" x 11", 5" tall. $2,000-2,500.

Beautiful Boulle work desk box with inlaid tortoise shell and tooled brass with bronze Doré trim. The inkwell and sander are covered with square cast Doré lids. The front is adorned with a medallion of a man's face. Winged angels surmount the four scroll feet. France, c. 1795. 10 3/4" x 14 1/2", 4" tall. $2,500-3,000.

Papier-mâchê—paper pulp mixed with glue then shaped, decorated, and lacquered—was popular in France in the eighteenth century and later in Germany and England.

Papier-mâché inkstand has a single crystal inkwell with a brass hinged lid. A stamp box in the center has a removable cover. The long channel across the front holds pens. Lacquered black with red and gold decorations. c. 1870. Inlaid with mother-of-pearl. 8" x 11". $300-400.

Papier-mâché inlaid with mother-of-pearl. Fitted with two swirl-pattern pressed glass inkwells with matching lids. Embellished with hand painted lilies of the valley and a gold border. Supported on four pad feet. England, c. 1880. 7" x 10". $275-350.

Papier-mâché inlaid with mother-of-pearl. Enhanced with red flowers and gold trim. Fitted with two crystal inkwells with faceted hinged lids with brass collars. England, c. 1880. 7 3/4" x 10 3/4". $250-350.

A stand made of gutta-percha, a molded resin. The single crystal inkwell sits in a frame on the base and has a hinged gutta-percha lid. A scroll design decorates the border. c. 1880. 5" x 6 3/4". $300-400.

Double inkstand made of gutta-percha, a molded resin. The crystal inkwells have gutta-percha lids with brass hinged collars. A classic design covers the flat base. France, c. 1880. 7 1/2" x 9 1/2". $450-650.

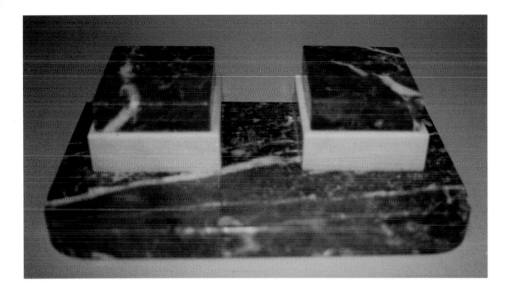

Flat based marble inkstand that has two square inkwells with loose lids. The channel across the front holds pens. c. 1920. 4 1/2" x 7 3/4", 2 1/2" tall. $175-250.

Large footed inkstand with two inkwells that have loose stacked covers. The long channel across the front is for pens. c. 1910. Marble. 7 3/4" x 13 3/8", 3 1/2" tall. $325-400.

Cut crystal inkwell with a hinged lid. The saucer shaped base has a scalloped edge with a pierced border. Marked on the bottom in cartouches "M. W. & Co. S. E. P. 83 4711 I.R." c. 1870. Silverplated. 7 3/4" in diameter, 4" tall. $200-300.

Ornate brass plated inkstand has a crystal inkwell with a ruby diamond decoration around the middle. The hinged lid has a cherub's face imposed on butterfly wings. The round saucer shaped base is enhanced with a decorative border and cherub faces. Marked "F. & Co. 953." Made by Elkington & Company. England, c. 1850. 8" x 10 1/2", 4 1/2" tall. $250-350.

A crystal inkwell with a hinged lid is seated in a square frame in the center of a wooden base. The backplate is supported on two columns and is graced with a lovely open grillwork design. Further enhancements are a shell and leaf decoration on the front and sides. Stands on four paw feet. England, c. 1900. Gilt brass. 5 5/8" x 8 1/4", 4 7/8" tall. $500-600.

Doré bronze inkstand with an array of pansies and leaves curved around the back and sides in an openwork design. The inkwell in the center has a hinged lid in the shape of a flower. Europe, c. 1880. 6" x 8", 3 1/2" tall. $500-700.

Lovely footed inkstand. Enameled portrait of a woman with long blond hair is framed in a circle on the top of the hinged lid. Shell shaped depressions are on the sides. Leaves and flowers in high relief add to the decoration. France, c. 1860. Bronze. 5" x 6 3/4", 1 3/4" tall. $700-1,000.

The head of a woman with wings is incorporated into scroll handles. The urn shaped inkwell has a hinged lid. The composition stands on six scroll feet. Europe, c. 1890. Gilt brass. 4 1/2" x 5 1/2", 5 1/2" tall. $450-600.

Tiny inkwell in the shape of a cap. The lid is hinged and opens to reveal an insert. France, c. 1900. Copper plated. 1 1/4" x 1 3/4", 1 1/8" tall. $95-125.

Beautiful Doré bronze inkstand. A cherub supporting a candle holder is standing on a base that is studded with cabochons of coral and jade. Two almandine crystal inkwells have engraved pagoda shaped covers. The front and back edges are rolled into a scroll. The composition stands on four ball feet. France, c. 1860. 4 1/4" x 8", 5 1/2" tall. $1,200-1,800.

Grapes and leaves form the base. A clear pressed glass inkwell is seated in the center and has a loose cover. America, c. 1900. Brass plated cast iron. 8" x 8 1/4". $200-250.

Inkwell on a pedestal that has a hinged lid with a ball finial. The saucer shaped base and the sides of the inkwell are set with red and green glass jewels. An engraved overall pattern adds to the decoration. c. 1900. Brass. 5" in diameter, 3 3/4" tall. $300-500.

A saucer with turned up edges. The glass inkwell has a brass band around the mid section. The lid is hinged. England, c. 1925. Brass. $125-175.

Ornately decorated inkstand with a pen rest across the front. The inkwell has a hinged lid with a finial. America, c. 1910. $200-250.

Brass stand with the figure of a lady standing in an open scrollwork backplate. The crystal inkwell has a loose lid with enameled flowers on top. The scalloped base stands on scroll feet. Probably English, c. 1885. 4 3/4" x 4 3/4", 4 1/4" tall. $250-350.

Globe shaped inkwell on a pedestal. The lid with a finial is hinged. On each side of the saucer base is a mythological animal with paw feet, a coiled tail, and a beak that is latched onto the pen rack. Europe, c. 1900. Brass. 6 1/2" x 9". $195-250.

Footed inkstand with a leaf and berry border. A pair of clear crystal inkwells with hinged lids are seated in the base. England, c. 1900. 4 1/2" x 7 3/8". $225-300.

Pretty inkstand with an enameled rose and green leaves on a trellis in the back. The recessed tray has two frames with wreaths on the front that hold the clear crystal inkwells. The lids are hinged. England, c. 1880. Brass. 4 3/4" x 9 3/4", 3 1/2" tall. $500-700.

Double stand with a lacy openwork rim. Two bulbous shaped crystal inkwells are fitted on a raised plateau on each side of the central handle. The loose covers are in the shape of flowers. Inscribed "Verlag Bei E. G. Zimmermann In Hanau." Germany, c. 1860. Cast iron with a bronze patina. 5 1/4" x 6", 3 1/2" tall. $300-400.

Highly decorated inkstand with an openwork backplate that has a face in the center. The base has four small posts to hold the two inkwells in place. The blue and white porcelain inkwells are grooved on the sides to fit between the posts. Marked on the bottom of the stand #1201. England, c. 1870. Inkwells: 2" x 2", 2 5/8" tall. Stand: 6" x 9", 6" tall. $600-900.

Footed tray with a ball on the back corners and a decorative border on the front and sides. The crystal inkwell is seated on a raised platform. A wreath adorns the front. The domed shaped lid is hinged. America, c. 1900. Brass. 6" x 9 1/2", 4" tall. $250-300.

Green leaves and purple grapes. The swirl glass inkwell has a loose cover that is encrusted with grapes and leaves. Marked #569. America, c. 1910. Cast iron. 5 1/2" x 5 1/2". $250-300.

Bronze inkwell with a hinged lid is attached to a black marble base. A sticker on the bottom informs us this item was "Made in France." 4 3/8" x 4 3/8", 2 3/4" tall. $150-250.

Pottery rolling inkwell in a bronze frame. The shell shaped saucer in the front is notched to make a pen rest. Embossed on the cross bar at the top: "Patent Perryian Gravitating Inkstand/Perry & Co. London." England, c. 1845. $300-400.

A paneled cut crystal inkwell has a round faceted lid with a hinged collar. The base is shaped like a leaf. c. 1900. Brass. 6" x 7", 4" tall. $150-250.

A square crystal inkwell is seated in a frame on a round base. The lid is hinged and the collar extends over the shoulder. The base is weighted and has a gadroon border. Europe, c. 1885. Brass. 3" in diameter, 2 1/2" tall. $125-175.

Square inkstand with handles. An elaborate design covers the entire frame. The cut crystal inkwell has a loose square cover. Stands on four scroll feet. Europe, c. 1885. Heavy brass. Base: 6 3/4" x 6 3/4". Well: 2 7/8" x 2 7/8". $275-350.

Small crystal inkwell that has a brass lid and collar that comes down over the shoulder. There is a cherub in high relief in the center of the hinged lid. America, c. 1900. 1 1/2" square, 2" tall. $95-150.

Two separate inkstands are joined to make one. Connected with a pen rack in the back and a square stamp box straddling the middle. The swirl glass inkwells are seated in frames on the base. The lids are hinged and have collars that extend over the shoulders. America, c. 1880. Stamped brass. 4" x 7 3/4", 3 1/4" tall. $250-350.

Art Nouveau stand with a milk glass inkwell fitted in a frame on the top of the irregular shaped tray. The hinged lid has a collar that extends over the top of the well. America, c. 1890. Brass. 3 1/4" x 6 3/8". $150-225.

Two milk glass swirl pattern inkwells sit in a recess on top of a footed tray. The lids are removable. Marked on the bottom of the wells "Anchor S. P. Co." America, c. 1900. Gold washed. 4" x 7 3/8", 3 1/4" tall. $250-350.

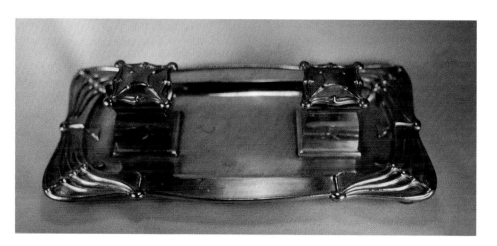

Flowing lines are on the ends of an oblong tray that holds two square inkwells. Decorative hinged lids. America, c. 1905. Brass. 5 1/2" x 11". $150-200.

Beautifully decorated urn that is supported on legs that have heads at the top and terminate in paw feet. The hinged lid has a flame finial. Germany, c. 1880. Brass. 4 1/8" in diameter, 6 1/4" tall. $400-600.

Heavily detailed single inkstand. Ribbons and swags connect the four lion heads around the sides. The legs terminate in paw feet. The beaded edged lid is hinged and has a lotus blossom finial. The glass insert is composed of two pieces: a funnel shaped insert fitted inside a regular insert. Probably England, c. 1860.
Bronze. 3 3/4" x 3 3/4", 4" tall. $500-700.

Heavily decorated footed stand that has two inkwells with hinged lids. Pen ledges are on all four sides. Stands on four paw feet. Marked in a diamond on the bottom "Bradley & Hubbard Mfg. Co." Meriden, Connecticut. America, c. 1900. Brass. 6 1/2" x 9 1/2". $250-350.

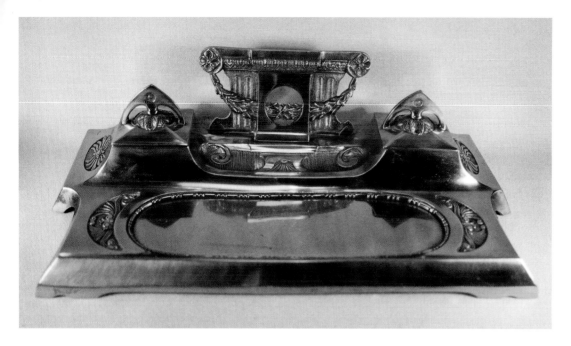

Art Deco inkstand with a draped swag across the columns in the back. The two inkwells have hinged lids with open-work finials. Between the inkwells is an oval depression for pen nibs. A large pen tray is across the front. England, c. 1910. Brass. 6 5/8" x 13 1/8", 3 1/2" tall. $400-600.

Inkwell in the shape of a rose with a stem. The top petal is hinged. c. 1900. White metal with a copper patina. 3 1/2" x 5", 2 5/8" tall. $200-275.

Palm fronds and branches cover the inkwell. The hinged lid is in the shape of a leaf. Marked on the bottom "23456 Tiffany Studio, New York." America, c. 1910. Bronze. 5 1/4" in diameter, 3 1/2" tall. $700-1,000.

Doré bronze Adams style oval inkwell decorated with a classic design. The lid is hinged. Inscribed on the bottom "Tiffany Studios, New York 1777." America, c. 1910. 3 1/4" x 4", 2 3/4" tall. $800-1,200.

Octagonal zodiac inkwell with a crab embossed on the hinged lid. Inscribed on the bottom "Tiffany Studios 842." America, c. 1910. Bronze. 2" x 4 1/4". $700-1,000.

Saucer shaped base with an openwork border and handles. The inkwell in the center has a hinged lid with a finial. The decorations include a child's face on the front of the inkwell, flowers, leaves, and scrolls. Europe, c. 1900. Bronze. 4 1/2" x 5 1/4", 3 1/2" tall. $125-175.

A typical Art Nouveau footed inkstand with graceful flowing lines and openwork borders. The inkwell has a hinged lid. Flowers decorate the base as well as the lid. Europe, c. 1900. Bronze. 4 1/4" x 5 1/4", 3 1/2" tall. $150-225.

Beautiful partner's inkstand that can be used from either side of a desk. The lids have ball finials and are hinged on the sides. There is a pen rest on the front and back. Finely detailed with ornate designs covering the entire composition. In the center front is a medallion with a man wearing a hat with wings. Scroll feet. c. 1890. Heavy brass. 7" x 9 3/4". $400-600.

Single partner's inkstand with a hinged lid that opens on the side. A medallion with a man in a winged hat is on the front. Pen rests are on all four sides. Beautifully decorated brass. This is a single version of the double partner's inkstand. 5 1/2" x 5 1/2", 3 1/4" tall. $300-400.

Inkwell on a pedestal. Cherries, leaves, stems, and small cup-shaped pods decorate the removable cover. Enhanced with a border of loops and pods. France, c. 1850. Bronze. 4" in diameter, 4 1/2" tall. $300-400.

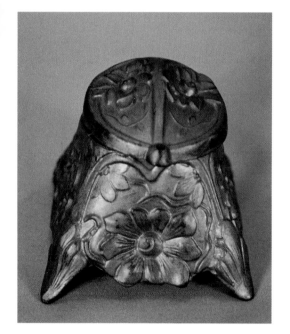

Small footed Art Nouveau inkwell covered with a floral pattern. The lid is hinged. America, c. 1900. Gilded white metal. 2 3/4" x 2 3/4", 2 1/4" tall. $100-175.

Art Nouveau double inkstand covered with leaves and berries. There is a blotter, or letter rack, in the back. The lids on the inkwells are hinged. Stands on four feet composed of leaves and berries. Inscribed on the back "G. F. 309. Deposé." France, c. 1890. 6 1/2" x 11", 5 3/4" tall. $750-900.

Square stand with side handles. An urn on a pedestal has a lion's head displayed on all four sides. The lid is hinged. The entire subject is covered in an ornate design. Stands on four scroll feet. Germany, c. 1900. Brass. 8 1/4" x 8 1/4", including handles 10". $200-350.

Oval stand with an inkwell on the right side and a sander on the left. The handle has a finger ring in the shape of a wreath. Decorated with a design of arches around the sides. France, c. 1890. (This is a copy of a 1790 inkstand.) Gilt bronze. 2 1/2" x 4 1/4", 5 1/2" tall. $900-1,200.

Large elaborately designed inkstand with an Oriental motif. On the four sides are face masks and fans. The hinged lid has the figure of Buddha for a finial. Enhanced with a beautiful reticulated border. Stands on a square footed base. England, c. 1885. Heavy brass. 9 1/8" x 9 1/8", 6" tall. $500-800.

A stand with four compartments: two have inserts, one has a nib cleaning brush, and one is for sealing wafers or pen nibs. The four lids are one piece and turn on a central axis to open. England, c. 1870. Brass and copper. 5" x 5", 4" tall. $1,000-1,500.

Open view of the four-compartment inkstand.

A leaf and scroll design covers this stand. The round lid with a finial is hinged. On the sides are four upturned ledges for pens. Stands on four three-toed paw feet. England, c. 1900. Brass. 5 3/4" x 7", 5 3/8" tall. $200-300.

Lovely stand with an openwork backplate. The two inkwells have hinged lids with finials. In the center is a three-compartment stamp box with a hinged lid. A scroll and leaf design covers the deep tray across the front. Marked "Germany." c. 1880. Heavy brass. 7" x 10", 5" tall. $300-450.

French Art Deco cloisonné inkwell with a hinged domed shaped lid. Red enamel over brass with "V R" in a diamond on the front. c. 1920. 2 1/2" x 4", 3 1/2" tall. $200-300.

Art Deco cloisonné inkwell. Black enamel over brass with a leaf design on the front. The domed lid is hinged. France, c. 1920. 2 5/8" x 4 1/8", 3 1/2" tall. $200-300.

The openwork backplate displays an urn at the top. There is an oval channel in the center for nibs. The area between the wells has a pierced leaf decoration with a beaded edge in front. The base is rounded on the ends and has claw feet. England, c. 1870. Brass. $300-350.

Inkwell made of lotus leaves with a flower finial. France, c. 1900. Brass. 3 1/4" in diameter, 2" tall. $100-150.

A design on the top of the hinged lid is the only ornamentation. America, c. 1920. Brass. 4 1/2" x 4 1/2", 2" tall. $85-125.

Scrolls and leaves cover the entire stand. The lid is hinged. Stands on foliated feet. Marked "G. L. 3 Deposé." France, c. 1890. 6 1/4" x 6 1/4", 3 1/2" tall. $225-275.

The scene on the top of the hinged lid depicts St. George slaying the dragon. Marked "Real Bronze." America, c. 1900. 6 3/4" in diameter, 4" tall. $350-500.

Highly decorated with cone shaped baskets of fruit on the corners and fleur-de-lis feet. France, c. 1900. Brass. 6 1/4" x 6 1/4", 3" tall. $250-300.

Single inkwell covered with leaves and scrolls. The square lid is hinged. Marked "D. L." and "Simone A. Raymond Le 16-10-29." France, October 16, 1929. Brass. $225-275.

Triangular shaped inkstand that is attached to a circular tray. The motif is faces. The inkstand's three legs have faces at the top of the leg and another on the knee. There are sunburst faces in between and the hinged lid has a face on the top. The legs terminate in claw feet. The tray has a floral border with a beaded edge. England, c. 1865. Brass. 9" in diameter, 4 1/4" tall. $250-350.

Oblong star-shaped stand with an openwork backplate and pen rack. The two pressed glass inkwells have removable covers. Marked "Pat' Apl'd For, Pat'd Nov. 12, 1878." America. Brass. 6" x 10 1/2", 7 3/8" tall. $250-350.

Framed with an ornate openwork border and backplate. The two inkwells have hinged lids. In the middle is a compartment for pen nibs or stamps. A pen rack is in the back. Inscribed on the bottom "Made In Deloium. J. H. No. 94." Germany, c. 1900. Brass. 8 5/8" x 15". $450-500.

A pair of urn shaped inkwells with handles are sitting on an elaborately decorated stand. A winged figure is on the backplate. There is a wide bracket in the rear that holds a large matching letter opener (not shown). England, c. 1880. Brass. 9" x 11". $600-800 for two-piece set.

Footed Art Deco stand with an openwork backplate. Six leaves decorate the front and three leaves cascade down on both sides. Hinged lid. America, c. 1925. Brass with an antique finish. 5 3/4" x 11", 3 1/4" tall. $250-350.

Chestnuts on a leaf. The large nut is the hinged lid. England, c. 1910. 5 1/2" in diameter, 2 1/2" tall. $125-150.

A leaf design is on the base and in a circle on the hinged lid. On the bottom is a mark for Bradley and Hubbard, Meriden, Connecticut. America, c. 1920. Brass. 5 1/2" x 5 1/2", 2 1/4" tall. $100-175.

Art Deco rolled brass with a pierced design on all four corners. Marked on the bottom "Austria." c. 1920. 4 3/4" x 4 3/4", 1 1/2" tall. $95-150.

An urn on a pedestal stands on a saucer shaped base. The lid is hinged. America, c. 1910. Brass. 7 3/8" in diameter, 6" tall. $125-200.

Urn shaped stand on a footed base with applied faces on each side. The lid is hinged and topped with a lyre finial. A ewer, musical instrument, and berries with leaves decorate the sides. A pen rest is on the base. Europe, c. 1900. Bronze. 2 5/8" in diameter, 5 1/4" tall. $250-350.

Art Nouveau stand with an amethyst glass inkwell with a loose lid. Lily pad shaped footed base with a pierced gallery around the back. France, c. 1900. White metal. 7" x 7", 4" tall. $175-225.

Art Nouveau footed inkstand with Egyptian style floral decorations. A pen rest is across the front. c. 1900. Bronze. 5 1/2" x 5 3/4", 3 3/4" tall. $195-250

Swirls and leaves cover this ornate stand. Details include a cherub's face in a bonnet, a dolphin's head, and at the bottom the head of a roaring lion. A shell shaped depression between the wells is for pen nibs. Germany or Austria, c. 1900. Iron with a bronze patina. 8" x 13". $200-300.

A long pen tray is in front of a square inkwell. The figure of a cherub is in the center of the tray. America, c. 1895. Brass. 7" x 11 1/4". $300-400.

A partner's inkstand that can be used on either side of a desk. The lids are hinged on the sides to open outward. An openwork centerpiece divides the stand in the middle and has hooks for a pen rack. c. 1860. Bronze. 9 1/2" x 13 1/2", 8" tall. $500-650.

Unique stand with a footed urn on a tray that has a pierced design on each end. The hinged lid has a fancy finial that is flanked by two snails. On the front of the urn is the head of a mythological beast with an open mouth. The handles form a pen rest. France, c. 1870. Bronze. 6 7/8" x 9", 6" tall. $600-900.

Elegant inkstand with many beautiful features. In the center is an attached candle holder with a snuffer, flanked by crystal inkwells with loose floral lids. A swan is on the front, a dog's head in the rear, and on each end the full figure of a pegasus. The base has openwork leafy feet. Europe, c. 1840. Brass. 7 1/2" x 12". $750-1,000.

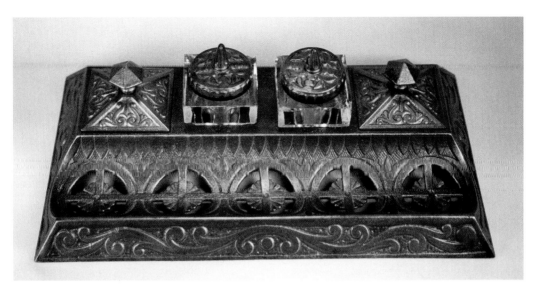

A four-compartment stand. A pair of square crystal inkwells are in the middle with a compartment on each side: one for stamps and the other for pen nibs. The lids are loose. The channel across the front is in an openwork design and holds pens. America, c. 1900. Cast iron. 5 1/4" x 9 1/2". $250-350.

The inkwell is set in an openwork brass frame of leaves and berries. A pen rack is attached to the onyx base. On the bottom is a mark with the inscription "Deposé #783." France, c. 1900. 5 1/2" x 6 1/2". $250-350.

Highly decorated stand with an inkwell in the center. The lid is hinged. All four sides have pen channels. c. 1900. Brass. 5 1/2" x 5 3/4", 4 1/2" tall. $200-250.

A lovely French inkstand with an attached candelabra. A pair of square crystal inkwells are set between brass posts on the base. The round faceted lids have a brass collar and are hinged. There is a depression for pen nibs between the wells. England, c. 1890. Brass. 6" x 9 3/4", 9 1/4" tall. $400-500.

Repoussé is a relief design that is created by hammering an object from the inside out. This inkwell has a very pretty floral pattern on the hinged lid as well as the sides. Derby Silver Co. #1735. America, c. 1895. Silverplate. 3" in diameter, 3 3/4" tall. $200-295.

The stand has handles and pen channels on the front and back. The inkwell has a hinged lid with a ball finial. Decorated with rosettes and scrolls. Supported on four paw feet. England, c. 1900. Brass. 5" x 7", 3 1/4" tall. $275-325.

Plain round inkwell with a hinged lid. Inscribed on the side "English Make." c. 1900. Brass. 4 1/2" in diameter, 2" tall. $95-125.

Lovely French inkstand that has a single urn shaped inkwell with a hinged lid. A wreath decorates the front and a swag is looped around the sides. There are handles on each side of the platform and a pen rest across the front. The composition stands on four feet. Europe, c. 1890. Brass mounts and frame on an onyx base. 6 1/2" x 10 1/2". $300-450.

Hut with a hinged thatched roof that opens to expose two milk glass inserts. There is a groove across the front for a pen. c. 1885. White metal. 3" x 4 1/2". $150-200.

Swirl glass inkwell encased in an openwork brass frame that is covered with grapes and leaves. The matching lid has a hinged collar. c. 1890. 3 1/8" x 3 1/8", 4" tall. $275-325.

This inkwell is part of a desk set that also has a letter stand, calendar holder, pen tray, rocker blotter, and four corners to fit on a desk blotter. America, c. 1910. Brass. $200-295 (inkwell only).

J. Loetz Witwe worked for Tiffany before establishing his own company in Austria in the late 1890s. Loetz pieces are seldom marked so positive identification is difficult. Iridescent art glass in shades of yellow, green, and blue with red streaks. Six sided with protruding bubbles. Mushroom shaped lid with a brass hinged collar. Austria, c. 1900. 3" in diameter, 5" tall. $900-1,200 (rare).

Beautiful inkstand that is part of a six-piece desk set. Not shown are a matching pen tray, nib brush, paperweight, matchbox holder, and seal. Extremely ornate inkstand with a grotesque mask in the center of the grillwork backplate. The two inkwells have hinged lids. Bun feet. Austria, c. 1885. Bronze. 7" x 14", 5 3/4" tall. $750-950 (inkwell only).

Loetz type iridescent art glass in shades of yellow, pink, and light green. Set in a square brass frame that has an openwork floral design. The brass leaf shaped lid is hinged. Austria, c. 1910. 4" x 4", 3 1/2" tall. $600-700.

Loetz type iridescent art glass in shades of green and gold with a bronze overlay on the shoulders. The lid is hinged. Austria, c. 1910. 3 1/4" x 3 1/4", 2 1/2" tall. $500-700.

Loetz type art glass with a brass hinged lid. Iridescent purple to green with black. Austria, c. 1910. 3 1/2" x 3 1/2", 2 1/4" tall. $500-700.

Loetz type green iridescent inkwell set in a pierced brass frame. Inscribed inside the hinged lid "G.E.R. 10385 D.R.G.M. 488430." Austria, c. 1910. 2 5/8" in diameter, 3" tall. $400-600.

Round Loetz type art glass with alternating bands of green and gold. The leaf shaped brass lid is hinged. Austria, c. 1910. 4 1/2" in diameter, 3 1/2" tall. $500-700.

Emerald green with bronze Doré ormolu mounts. The hinged lid has a decorative wreath on the top. France, c. 1900. 2 1/2" square, 3 1/2" tall. $700-1,000.

Loetz type cranberry art glass. The inkwell is encased in an openwork frame. The lid is hinged and displays a grotesque mask with an open mouth and ram's horns. Austria, c. 1900. Brass mounts. 3 1/2" x 3 1/2", 3" tall. $500-700.

Loetz type cranberry glass in the shape of a drum encased in an openwork frame. The brass lid is hinged and has a grotesque face on the top with an mouth and ram's horns. Marked under the lid "0385 D.R. 168180." Austria, c. 1900. 2 5/8" in diameter, 2 3/4" tall. $500-700.

Round Bristol blue inkwell with an old Sheffield plate hinged lid. England, c. 1790. 4 1/2" in diameter, 2 1/2" tall. $700-1,000.

Laminated green and white swirl pattern favrile glass. A small paper label reads "Tiffany Registered Trade Mark #189T." America, c. 1910. 4" in diameter, 2 1/4" tall. $700-1,000.

Porcelain with flowers in frames around the sides. The hinged lid and footed base are sterling silver and are covered with an ornate design of leaves and scrolls. Marked with a blue crown with a D underneath. Dresden. c. 1870. $600-900.

Ribbed blue glass. The brass lid is hinged and is inscribed on the top "Encrier De Voyage F.C.J. ne. A Paris Brevete." France, c. 1880. $400-600.

Thomas Webb and Sons, established in 1837 in Stourbridge, England, are noted for their cameo glass. This inkwell has a white leaf design on frosted amber with an engraved sterling silver mushroom shaped lid. 3 1/2" in diameter, 3 3/4" tall. $2,500-3,500 (rare).

Wave crest is the trade name used by the C. F. Monroe Co. of Meriden, Connecticut. Opaque blown in the mold glass blanks were purchased from Pairpoint Mfg. Co. and decorated with floral designs. This inkwell has a sterling silver hinged lid. America, c. 1892. 3 1/4" in diameter, 2 3/4" tall. $300-400.

Multi-colored hand blown glass inkwell with a double ball stopper. France, c. 1910. 3 1/2" in diameter, 4 3/8" tall. $300-400.

Paperweight partner's inkwell. The lid is hinged in two directions so two people can use it. Red and white glass in a swirl design with a brass pineapple lid and collar. Marked on the rim "Pat. 1-10-49 Betjamanns." England. $600-900.

Art Nouveau Loetz style stand in the shape of a lily pad. The openwork backplate is a pen rack. The cranberry glass inkwell has a leaf shaped hinged lid. Austria, c. 1900. Nickel-plated over iron. 6 1/2" x 10", 3" tall. $500-700.

Red paperweight partner's inkwell. The brass dome-shaped lid is hinged in two directions. Marked "Betjamanns Patent 11040." England, c. 1880. 4 1/2" in diameter, 5 1/2" tall. $600-900.

Jet-black glass with a painted floral design on the front and the top. The faceted lid is removable. America, c. 1900. 2 1/2" square, 3" tall. $250-300.

Jet-black glass with a painted desert scene on the front. Faceted hinged lid with a brass collar. America, c. 1900. 1 1/2" square, 2 1/8" tall. $150-250.

Mount Washington square lusterless glass inkwell in a swirl pattern. The matching round lid has a hinged brass collar. America, c. 1890. 2 1/2" square, 3 1/2" tall. $250-350.

Mount Washington lusterless pink glass in a swirl pattern. Decorated with hand painted flowers. The matching round lid has a hinged brass collar. America, c. 1900. 2" in diameter, 2 7/8" tall. $300-400.

Bronze stand with a pair of cranberry inkwells with loose caps. One bottle is for ink and the other is a sander. The stag's horns form a pen rest. England, c. 1900. 4 3/8" x 7 1/4", 3 1/2" tall. $300-400.

Mount Washington square lusterless blue glass in a swirl pattern. Decorated with hand painted white daisies. The matching round lid has a brass collar with a hinge. America, c. 1900. 2 1/8" in diameter, 3 3/8" tall. $250-350.

Tea kettle style inkwell. Venetian glass cranberry to clear accented with gold. The hinged lid on the spout is made of brass. On the top is a brass disc with a crown inscribed "Mordan & Co. London." England, c. 1850. $1,000-1,500 (rare).

Loetz type art glass with a hinged brass lid. The white base is criss crossed with red streamers. Austria, c. 1910. 4" in diameter, 1 1/2" tall. $700-900.

Quilted light blue and turquoise satin glass with a brass hinged lid. America, c. 1880. 3 1/8" square, 4" tall. $600-800.

Ruby glass with a white overlay forming round circles. The silverplated cover is hinged and has a floral decoration on the top. The shell in the front holds pen nibs. A pen rack in the back holds three pens. c. 1870. 4" in diameter, 5 3/8" tall. $900-1200.

Paperweight inkwell with red to clear glass. The faceted lid has a brass hinged collar. c. 1900. 4 1/2" in diameter, 3" tall. $300-400.

Bohemian glass red to clear overlay. The matching top has a hinged brass collar. Scalloped base. c. 1860. 5" in diameter, 4" tall. $800-1,200.

Ruby glass inkwell with a matching lid. The brass collar is hinged. Inscribed on the bottom "Aug. 27, 1895. Made in U. S. A." $400-600.

Blue pressed glass daisy and button pattern with a matching loose cover. America, c. 1900. 2" x 2", 2" tall. $250-350.

Brilliant blue octagonal cut crystal. The hinged lid has a brass collar. c. 1885. $400-550.

Tiny bright blue cut crystal. The matching lid is hinged with a silverplated collar. America, c. 1890. 1 1/8" x 1 1/8", 1 3/4" tall. $300-400.

Intaglio cut flowers are on four sides of this octagonal shaped inkwell as well as the faceted lid. Brilliant blue cut crystal with brass mounts. c. 1885. 2" in diameter, 3" tall. $450-600.

Brilliant blue cut crystal. Pagoda shaped with a brass hinged collar. c. 1885. 2" x 2", 3 1/2" tall. $500-650.

Brilliant blue cut crystal pyramid with a hinged lid. c. 1890. 2 1/8" x 2 1/8", 3 1/2" tall. $350-450.

Double blue glass with a pen tray across the front. The inkwells have brass hinged lids with a sliding cover over the dipping holes. c. 1900. $400-500.

Ball pyramid in brilliant blue crystal. The matching lid with a ball finial has a brass hinged collar. Painted flowers are added decorations. France, c. 1880. $600-800.

Brilliant blue crystal with cut corners. The faceted lids on the two wells have brass hinged collars. America, c. 1885. 2 1/8" x 3 1/2", 2 5/8" tall. $400-600.

Brilliant blue pyramid shaped cut crystal. The faceted lid has a brass collar and hinge. America, c. 1885. 3 1/2" x 3 1/2", 3" tall. $400-500.

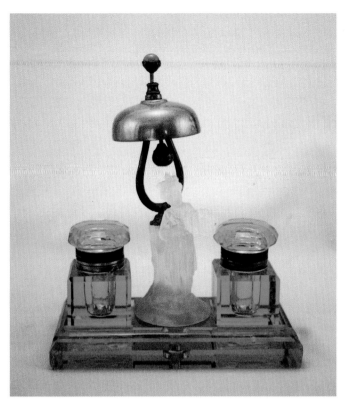

Unique stand with two bright blue cut crystal inkwells. The faceted hinged lids have brass collars. A frosted glass figure of a woman is standing in the middle. A calling bell with a blue glass finial is in the back. c. 1890. 2 5/8" x 5 3/8", 6 5/8" tall. $800-1000.

Brilliant blue cut crystal with a Mary Gregory type painting on the front. The faceted hinged lid has a brass collar. Mary Gregory worked in the decorating department of the Boston and Sandwich Glass Co. in Sandwich, Massachusetts. c. 1880. 2" x 2", 3 1/4" tall. $400-600.

Brilliant blue cut crystal. The faceted lid is hinged and has a brass collar. c. 1890. 2 3/8" x 2 3/8". $350-500.

A pair of sparkling blue cut crystal inkwells. Brass hinged mounts. c. 1890. 1 3/8" x 1 3/8". $275-350 each.

Brilliant blue cut crystal with hinged brass mounts. Faceted mushroom shaped lid. Three steps for pens. c. 1890. 1 3/4" x 3 1/2", 2 3/4" tall. $400-550.

Brilliant blue crystal with hinged brass mounts. The base has three ledges to accommodate pens. c. 1890. 2 5/8" x 2 7/8", 2 3/4" tall. $400-550.

Brilliant blue diamond-shaped tray with a clear crystal inkwell. The faceted lid has a brass collar that is hinged. c. 1885. 4 3/4" x 6 1/8", 3 1/2" tall. $350-500.

Small clear crystal inkwell with a brilliant blue faceted lid and a hinged brass collar. America, c. 1885.
1 1/2" x 1 1/2", 2 1/2" tall. $195-295.

Clear crystal with a brilliant blue faceted lid. The brass collar is hinged. c. 1885.
2 1/4" x 2 1/4", 3" tall. $175-275.

Amber cut crystal inkwell that has a matching faceted lid. The brass collar is hinged. Cut channels around the base hold pens. America, c. 1910.
2 3/8" x 2 3/8", 4 1/2" tall. $600-800.

Amber cut crystal. The faceted hinged lid has a brass collar. c. 1900. 2 3/8" x 2 3/8", 3 1/2" tall. $300-450.

Amber cut crystal with two inkwells. The faceted lids have brass hinged collars. America, c. 1885. 1 3/4" x 3 1/2", 2 5/8" tall. $600-800.

The Saint-Ann glassworks was established in Baccarat, France, in 1764, and is still in business today. They are noted for their fine glassware and paperweights. This Baccarat glass inkwell is shaded from green to clear in a swirl pattern. The hinged lid has a brass collar. France, c. 1880. 3" square, 4 1/4" tall. $400-600.

Daisy and button Vaseline glass. Matching loose lid. Believed to be Sandwich glass. America, c. 1900. 2" x 2", 2" tall. $250-350.

Emerald green crystal. Typical Art Nouveau silver overlay with flowing leaves and flowers. A faceted diamond band encircles the mid section. Marked 800 on the front of the collar, which indicates this is continental silver. Matching hinged lid with silver mounts. Germany, c. 1910. $700-900.

Emerald green glass horseshoe trimmed in gold. Inscribed on the back "Patent Apd. For." America, c. 1900. 3 3/8" x 4", 2 1/4" tall. $250-400.

Inkwell with a green flower in the bottom that radiates color through the cut crystal prisms. Sterling silver top. Made in Birmingham, England, c. 1890. 3 3/4" in diameter, 3 3/8" tall. $600-800.

Hand blown tea kettle type inkwell, green glass, with a pontil scar on the bottom. A pontil point was left when the punty, or blowing rod, was removed from the blown piece. America, c. 1860. 2 1/4" x 3 3/4", 2" tall. $200-300.

Green to clear glass with a matching domed lid. The brass collar is hinged. America, c. 1900. 3 1/2" in diameter, 5" tall. $275-325.

Green glass with a cut jewel set in the brass lid. c. 1900. 3" in diameter, 2 1/8" tall. $400-600.

Art Nouveau stand with green stained glass under a brass grid frame. The clear crystal inkwells have matching stained glass hinged lids. Stylized flowers decorate the lids and the sides of the stand. America, c. 1900. 4 5/8" x 8 1/2", 3 1/4" tall. $600-800.

Unmarked Lalique style frosted glass standing on a black glass base. The hinged lid has the figure of a little girl reading a book on the top. France, c. 1905. Brass hinged collar. 4 5/8" in diameter, 5 1/2" tall. $400-600. Rene Lalique (1860-1945) was a prominent jewelry maker. About 1900, he began making glassware. He produced many pieces in the Art Noveau and Art Deco style.

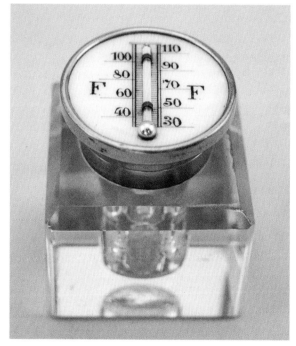

A thermometer is set in the top of the crystal inkwell. The nickel-plated brass lid is hinged. England, c. 1910. 2 1/4" x 2 1/4", 2 5/8" tall. $400-600.

Watch stands were sold without the watch so a purchaser could use his own. When a gentleman went to his office, he would take the watch out of his watch pocket and suspend it on the inkwell thus forming a clock. At the end of the day he would remove it from the stand and put it back in his pocket. The sterling silver lid on this crystal inkwell is engraved with a snail and berries. The lid is hinged and opens to display a watch encased under the lid. Hallmark for Birmingham, England, 1900. 3 3/4" x 3 3/4", 4" tall. $900-1,200.

Open view of inkwell showing the watch under the lid.

Six-sided cut crystal with a sterling silver watch stand on the lid. Marked "Mappin & Webb 9223." Hallmark for Birmingham, England, 1914. 3" x 3", 4" tall. $1,500-1,800.

Sterling silver inkstand with a watch in a decorative frame in the back. White overlay on a clear glass. The inkwell and sander have hinged lids with ornate finials. The shell shaped tray in the front is for accessories. Stands on four ball feet. America, c. 1890. $1,000-1,500.

Eight-day Swiss made pull-out clock in the shape of a coal hod. The inkwell is under the hinged lid. There are two glass inserts inside. Sterling silver with a hallmark for Birmingham, England, 1910. 1 5/8" x 2 5/8", 3" tall. $1,000-1,500.

Crystal inkwell with a hinged lid, fitted between posts on a twig shaped frame. Affixed to the branch on the right side is a pink glass rose bud with a pale green leaf. c. 1870. Brass. Inkwell: 2 1/8" square. Frame: 4 7/8" x 5". $300-450.

Leaves and flowers are intertwined on a tree branch. A crystal inkwell with a hinged lid and a brass collar is fitted between posts on the base. c. 1870. Brass stand. 4 1/2" wide, 5" tall. $300-400.

Layered cut crystal. The hinged lid has the portrait of a woman set in a beaded bezel. America, c. 1885. Silverplated lid. 2 1/2" in diameter, 3 1/4" tall. $225-300.

Art Nouveau clear crystal inkwell. The brass lid has a floral border that frames a portrait of a young woman. England, c. 1900. 3 1/4" in diameter, 3 1/4" tall. $400-500.

Pressed glass with a tray in the front and a brass hinged collar. c. 1900. 4 1/8" x 4 1/4", x 3 1/2" tall. $100-150.

Cut crystal with hand painted flowers on frosted medallions around the sides. The faceted lid has a flower on the top. Hinged brass collar. c. 1900. 2 1/2" square, 3 1/4" tall. $200-300.

Hand blown and cut crystal with a yellow goblet inside. The separate stopper lid has a knob on the top. This inkwell is seldom found intact; the loose lid is often missing. Octagonal shape. France, c. 1900. 3 1/4" tall. $1,000-1,200.

Millefiori (thousand flowers) is a type of mosaic glassware. The technique for making this colorful glass was invented by ancient Egyptian craftsmen in first century B.C. Venetian glassmakers refined the process in the fifteenth century. This art form had a revival starting in the late 1800s. Pictured is a paperweight inkwell with a matching mushroom shaped lid. Brilliantly colored in shades of red white and blue. c. 1900. 4 1/2" in diameter, 5 1/2" tall. $800-1000.

Baccarat inkwell with bulbous iris swirls. The sterling silver lid with embossed roses was made by Gorham Silver Co. Engraved on the rim "Sterling 790." c. 1900. 4" in diameter, 3 1/2" tall. $1,200-1,400.

Clear crystal inkwell with a hinged lid. The bottom is cut in a cane pattern and the design reflects upward through the crystal creating a pretty effect. The lid is sterling silver with hallmarks on the rim that tell us this piece was made in Birmingham, England, in 1903. 3 1/2" x 3 1/2", 2 1/2" tall. $200-300.

Large glass inkwell in a swirl pattern. The lid is hinged with a nickel-plated brass collar. It has a hollow bottom instead of the more common solid base. America, c. 1900. 3 3/4" x 3 3/4", 5 3/4" tall. $200-300.

Square crystal inkwell with a loose brass cover. Inscribed on the front in gold letters "Carter's" (Carter's ink). America, c. 1928. 4" x 4", 3 3/4" tall. $150-250.

Art Nouveau clear crystal. The sterling silver lid has a floral border that frames a portrait of a woman with long flowing hair. Inscribed "B. & F. Sterling 1069." America, c. 1900. 2 1/2" in diameter, 2 7/8" tall. $350-500.

Round cut crystal. The sterling silver lid is hinged and has a floral border. America, c. 1900. 5" in diameter, 3" tall. $500-700.

241

Clear crystal with two gadroon bands around the sides. The silver lid is hinged and has a beaded edge around the top. Marked on the bottom "Crystal. St. Louis, France." 3" in diameter, 3 1/4" tall. $350-500.

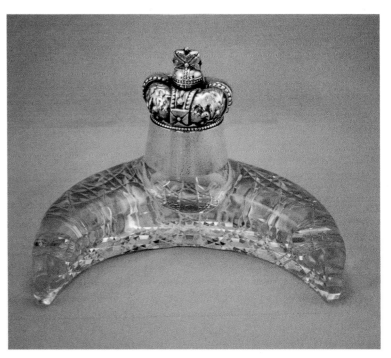

Clear crystal crescent shaped with a silverplated lid in the shape of a crown. There is an indentation on each tip to hold a pen. England, c. 1880. 5 1/2" long, 3 3/4" tall. $250-350.

Footed cut crystal with a hinged sterling silver lid. A band with faces between apples is around the lid. Inscribed on the rim "Gorham Sterling, D 435." America, c. 1890. 3 1/2" in diameter, 4" tall. $500-700.

Crystal with bulbous shaped swirls. The hinged lid is sterling silver. Probably Baccarat. France, c. 1900. 4" in diameter, 4 1/4" tall. $350-500.

Art Nouveau clear crystal inkwell. The hinged lid is sterling silver and displays a cherub astride a seashell. The bulbous well is in the shape of a scallop. America, c. 1900. 3 1/2" in diameter, 3 1/2" tall. $300-450.

Swirl glass bottom with a sterling silver hinged lid. Marked on the collar with a Gorham Silver Co. hallmark, which is a lion, anchor, a G, and "Sterling 53 1/3." A date letter tells us this was made in 1869. America. 3 3/8" x 3 3/8", 4 1/2" tall. $500-600.

Clear crystal scallop shaped inkwell with a brass hinged lid. America, c. 1900. 4" in diameter, 3" tall. $200-250.

A glass waffle-patterned inkwell shaped like a fish. The hinged cap is made of nickel. c. 1910. 2 1/2" x 5". $200-250.

Large hexagonal cut crystal inkwell. France, probably Baccarat. The round matching cover is removable. 5 1/2" x 6 1/4", 5 3/4" tall. $500-700.

Cut crystal with a mushroom shaped hinged lid. England, c. 1900. 2 1/4" x 2 1/4", 3 1/4" tall. $175-225.

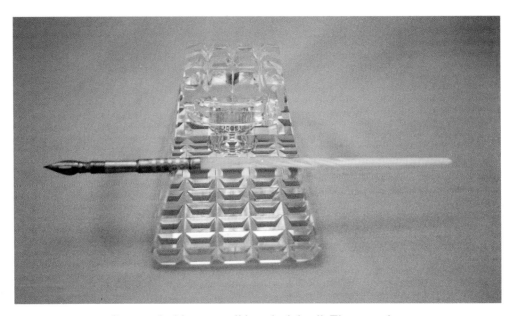

Cut crystal with a square lid on the inkwell. The stepped pyramid has a channel across the front that holds a pen. 2 5/8" x 4 7/8", 2 5/8" tall. $150-200.

Shield shaped crystal with a sterling silver hinged lid. A channel across the front holds a pen. There is a hallmark on the lid for Birmingham, England, 1927. 2 3/8" x 3 1/2", 1 1/2" tall. $200-300.

Square crystal with a dark blue enamel and brass lid. In the center of the hinged lid is a round crystal disc set in a brass bezel. c. 1900. 2" x 2", 2 1/4" tall. $225-325.

Triangular cut crystal inkwell with a hinged silver lid. c. 1900. 4 7/8" x 4 7/8", 2 3/8" tall. $250-350.

Glass railroad inkwell with a hinged metal lid. Inscribed on the top are an S, a crown, an O, and 44-12. England, c. 1920. 2" x 2", 2" tall. $75-95.

Pressed glass with a brass hinged lid. Marked "Made in England" on the bottom. c. 1900. 2 1/2" x 2 1/2", 2 1/2" tall. $75-100.

Crystal inkwell with a matching hinged lid. Nickel-plated brass collar. America, c. 1900. 2 1/8" x 2 1/8", 2 7/8" tall. $125-175.

Controlled bubble set. A sponge pot is on the left and an inkwell on the right. Pairpoint glass marked "Goodnough & Jenks." Covered with loose sterling silver lids. America, c. 1890. The sponge pot is 3" in diameter, 2 3/8" tall. The inkwell is 2 3/4" in diameter, 3 1/4" tall. $600-800 for set.

Crystal in the form of a house made of bricks. The pyramid shaped lid is on a spindle and slides to one side to open. England, c. 1900. 2 1/8" x 2 1/8", 2 5/8" tall. $150-250.

Paneled bulbous shaped inkwell. The brass lid is hinged and has a floral design on the top. England, c. 1900. 4" in diameter, 2 1/2" tall. $125-200.

Square crystal inkwell with a ribbed brass cover. England, c. 1900. $100-150.

Square cut-crystal inkwell with a hinged lid. Marked "Shreve & Co. San Francisco. Sterling & Other Metals." c. 1900. 2 1/2" x 2 1/2", 3" tall. $175-250.

Tiny cut crystal inkwell with a sterling silver hinged lid. A decorative border encircles the top. America, c. 1900. 1 3/8" x 1 3/8", 2" tall. $150-250.

Scalloped clear glass with a brass cover. The knob in the center controls the opening and closing of the two ink holes. One side is for "Red" and the other for "Black." The embossed message reads "E. Edwards Manufacturer. Birmingham" and "The Excelsior Inkstand. Registered Oct. 7th, 1851." England. 4 5/8" in diameter, 2 7/8" tall. $350-450.

Square crystal with a waffle pattern on the bottom. The silver lid screws on. Flowers decorate the top. c. 1900. 1 1/2" x 1 1/2", 1 7/8" tall. $100-150.

Six-sided crystal inkwell with a sterling silver hinged lid. c. 1900. 1 7/8" x 1 7/8", 2 1/4" tall. $125-195.

Cut crystal with a hinged silver lid. A floral and beaded design encircles the top. c. 1900. $125-195.

Square crystal with cut corners. The silver lid has a floral design on the top and is encircled with beads around the edge. c. 1900. 1 7/8" x 1 7/8", 2 1/4" tall. $95-150.

Glass paperweight inkwell with a brass hinged lid. England, c. 1885. 3" in diameter, 1 1/2" tall. $125-200.

Faceted crystal inkwell. The eight-sided black glass lid has a brass collar that is hinged. France, c. 1920. 1 1/4" x 1 1/4", 2" tall. 100-150.

Crystal inkwell with a hinged brass collar is attached to a saucer shaped base. England, c. 1930. Brass. $100-175.

Round blown in the mold glass inkwell with a hinged brass lid. c. 1890. 3 1/8" in diameter, 1 5/8" tall. $125-150.

Round glass with a metal lid that has a flower in the center. The lid is hinged. America, c. 1910. 3 3/4" in diameter, 2 1/2" tall. $75-125.

Crystal with a brass dome shaped hinged lid. America, c. 1900. 2 1/4" x 2 1/4", 3" tall. $150-225.

Square cut crystal. The ball shaped hinged lid is sterling silver. America, c. 1900. 3" x 3", 5" tall. $200-250.

A Victorian pyramid of glass balls. The inkwell has a lid with a hinged brass collar. England, c. 1880. 3" in diameter, 5" tall. $200-300

A large pyramid of iridescent glass balls. The matching lid has a ball finial and a hinged brass collar. England, 1880. 4 1/4" x 4 1/4", 6" tall. $200-300.

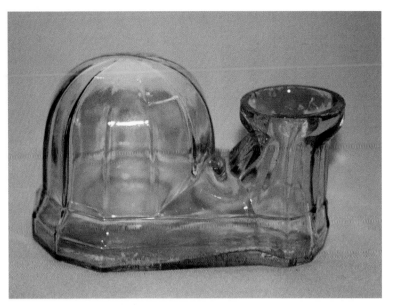

Clear glass gravity fed inkwell. Inscribed "Morgan's Patent July 16, 1867." America. 3" x 5", 3" tall. $85-125.

Gravity fed sunglass inkwell. The chemicals used in this glass turns it from clear to amethyst when exposed to sunlight. Inscribed on the bottom "Little." America, c. 1875. 2 1/2" x 4", 2 1/2" tall. $85-125.

Tiny pressed glass amethyst inkwell in a swirl pattern. The matching round cover is removable. America, c. 1905. 1 1/2" x 1 1/2", 1 3/4" tall. $100-150.

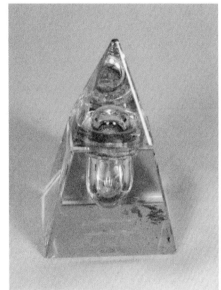

Cut crystal pyramid with a hinged lid. c. 1880. 1 5/8" x 1 5/8", 2 3/4" tall. $200-300.

Tiny green pressed glass inkwell in a swirl pattern. The matching round lid is removable. America, c. 1905. 1 1/2" x 1 1/2", 1 3/4" tall. $100-150.

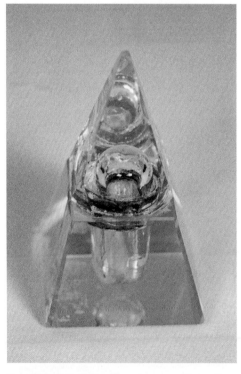

Pyramid shaped Vaseline glass (a yellow, oily appearing glass). The lid is hinged. c. 1880. 1 5/8" x 1 5/8", 3" tall. $400-500.

Glass with a screw on brass lid. In a circle on the top is a crown with the message "Mosley's Revolving." This inkwell comes from a writing box or lap desk. England, c. 1900. 1 3/4" 1 3/4", 1 5/8" tall. $75-100.

A tiny swirl glass inkwell that has a hinged lid with a brass collar. c. 1885. 1" x 1", 1 3/4" tall. $100-150.

Square crystal with cut edges. The hinged lid is inscribed "E.P.N.S.," which stands for electro-plated nickel silver. On the top is a pair of enameled red, white, and blue flags and a banner that reads "Saxon." England, c. 1900. 1 3/4" x 1 3/4", 1 3/4" tall. $150-250.

Square crystal with a brass lid that is hinged. England, c. 1900. 3" x 3", 3" tall. $100-175.

Cut crystal with a brass mushroom-shaped hinged lid. England, c. 1900. 2 3/8" x 2 3/8", 3 1/2" tall. $100-175.

Paneled star cut crystal. The brass mushroom shaped lid is hinged. England, c. 1900. 2 1/4" x 2 1/4", 3 1/4" tall. $100-175.

Pressed glass with a brass mushroom-shaped hinged lid. 2 1/2" x 2 1/2", 3" tall. England, c. 1900. $100-150.

Hexagonal heavy art glass. Intaglio cut swags pinned up with bows encircles the sides. Matching loose faceted cover. c. 1910. 3 1/2" x 3 1/2", 4 1/2" tall. $350-450.

Hexagonal cut crystal with a matching faceted lid. Hinged brass collar. c. 1860. 3 1/4" x 3 1/4", 3 1/2" tall. $150-250.

Cut crystal with a matching faceted lid. Brass hinged collar. England, c. 1900. 3" x 3", 3 3/4" tall. $300-400.

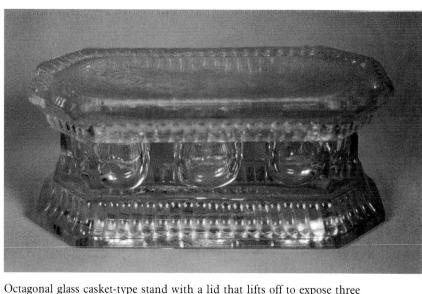

Octagonal glass casket-type stand with a lid that lifts off to expose three inkwells: one for red ink, one for blue, and one for black. America, c. 1890. 2 5/8" x 5 3/8", 2 3/8" tall. $150-200.

Cut crystal with a matching loose cover. c. 1920. Base: 2 1/2" x 2 1/2", 3 1/4" tall. $300-400.

Clear glass with a copper band over the top to hold the loose lid in place. There are two dip holes in the top: one side says "Shallow," and the other side says "Deep." The lid rotates the opening from one side to the other. Marked "Pat'd Jan. 5, 1904." America. 4 7/8" in diameter, 3 1/2" tall. $175-225.

Very small swirl glass inkwell. The matching lid with a brass collar is hinged. c. 1890. 1 1/8" in diameter, 2" tall. $50-75.

Square crystal. The hinged mushroom shaped lid is silverplated. America, c. 1900. 1 3/4" x 1 3/4", 2 1/2" tall. $75-100.

Clear swirl glass. The matching dome shaped lid has a brass collar and is hinged. America, c. 1900. 1 1/2" x 1 1/2", 2 3/8" tall. $75-125.

Clear crystal in a swirl pattern. The sterling silver lid has a hallmark for Birmingham, England, and a date letter for 1891. This inkwell came from an inkstand and would be one of a pair. 1 3/4" in diameter, 2 3/4" tall. $125-200.

Small diamond shaped cut crystal inkwell that stands on four feet. The matching faceted lid has a hinged brass collar. America, c. 1920. 1 3/4" x 2 1/4", 2 1/2" tall. $75-100.

Small square crystal inkwell. The round faceted lid has a brass hinged collar. America, c. 1920. 1 1/2" x 1 1/2", 2" tall. $75-95.

Round pressed-glass in a diamond pattern. The silverplated top has a dip hole in the center and is hinged. This type of inkwell fits in an inkstand. England, c. 1850. 1 1/2" in diameter, 2 1/2" tall. $95-125.

Square crystal with cut corners. Separate cast brass cover. America, c. 1895. 2" x 2", 2" tall. $75-100.

Hand blown iridescent glass with a ground bottom. The brass collar is hinged. America, c. 1885. 3 1/4" in diameter, 4" tall. $200-300.

Pressed glass stand with three inkwells. The hinged mushroom shaped lids are silverplated. Channeled around the front and sides for pens. America, c. 1900. 3 1/2" x 10", 3 1/4" tall. $200-300.

Clear glass with two inkwells that have loose bakelite covers: one red and one black, to indicate the color of ink. Three depressions in the front for pen nibs. A pen rest is across the top. America, c. 1925. $125-175.

Clear glass stand with two inkwells that have cast iron swivel covers. The pen rest on the front hold two pens. "Tatum" is embossed on the bottom. Tatum Manufacturing, Cincinnati, Ohio. America, c. 1885. 3" x 6", 2 3/4" tall. $200-300.

Clear glass inkwell with a bakelite lid that slides to open. A pen rest is across the front. America, c. 1910. 3 1/2" x 5", 2 3/4" tall. $75-125.

Clear glass with a bakelite sliding pen tray on the top. Embossed on the glass "Slide On." Under the lid is a V in a diamond with "Velds" spelled out on top. Inscribed under the lid "Velos Series." America, c. 1900. 3 1/4" x 3 1/4". $100-175.

Round glass with a separate bakelite lid. Written on the top is "SENGBUSCH U. S. A." Self-closing. 3" in diameter. $45-65.

Square glass. Written on the lid "Sengbusch Self Closing Inkstand Co. Milwaukee, Wis. Pat. Apr. 21, 03. Aug. 23, 04." Bakelite in three pieces: the inkwell insert, rim, and lid. 3" x 3". $65-95.

Art Deco black glass with a pair of clear crystal inkwells fitted in brass frames. The black glass lids have brass collars and are hinged. A pen rest is across the front. America, c. 1925. 5 1/8" x 8 3/4", 3 1/4" tall. $200-300.

Art Deco inkwell. The square black glass lid is hinged. England, c. 1920. 2" x 2", 2 1/2" tall. $75-95.

Clear glass with a black glass funnel-shaped insert. There is a pen channel across the top and another one across the bottom. The bottle can be reversed top to bottom. c. 1930. 2 1/2" x 2 1/2", 2" tall. $75-125.

This heavy blown-glass inkwell has a finial and a rough pontil scar on the bottom. The open dip cup on the side provides access to the ink. America, c. 1850. 3 3/4" in diameter, 3" tall. $150-250.

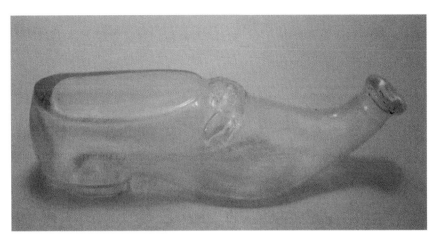

Blown ink bottle in the shape of a glass shoe with a pointed toe. America, c. 1855. 1 1/4" x 6", 1 1/2" tall. $100-150.

Brass inkwell on a round glass base. The lid is spring loaded. On the edge of the base is engraved "England Make." c. 1920. 4 1/4" in diameter, 2 1/8" tall. $150-200.

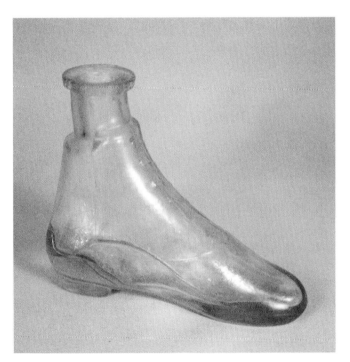

Blown glass ink bottle in the shape of a high top shoe. America, c. 1855. 4" long, 3 3/8" tall. $100-150.

Pair of square opalescent glass inkwells fitted in round bases on a flat tray. The hinged lids extend over the shoulders of the wells. Pressed brass. c. 1885. 4 1/2" x 6 1/2". $100-150.

Art Nouveau stand with a clear crystal inkwell. The mushroom shaped lid is hinged. The paw-footed tray has hibiscus on each end. France, c. 1890. Silverplate. $175-225.

A beautiful openwork brass frame holds a square crystal inkwell with a loose cover. France, c. 1885. 4 7/8" x 4 7/8", 4 1/8" tall. The well is 2 3/8" square. $250-350.

Unusual inkstand with four clear glass inkwells: two on the sides, one in the back, and a small one in the middle. The small one has a glass funnel insert and a glass stopper. All have loose covers. A pen rack is across the front. America, c. 1900. 5" x 6", 3 1/4" tall. $150-250.

A baby, bottom side up, is resting on the top of a round open inkwell. Marked on the bottom "Denmark. D. 1704" and in a diamond "Just." Bronze. 3" total width, 2 3/8" tall. $200-300.

Folk art inkwell that is hand carved from a piece of coal. The pyramid-shaped lids are hinged. On the front are hearts carved in a square block. America, c. 1910. 2 1/8" x 3 1/4", 2 3/4" tall. $100-150.

Although inkwells were sometimes attached to electric lamps, we believe the torch held up by the boy on this inkstand once held a candle and it was electrified at a later date. The inkwells have hinged lids. America, c. 1920. Patinated white metal. $250-300.

Large stand with two inkwells that have hinged lids. The letter holder in back has an openwork design with a cherub on the top and one on each side holding a basket of flowers. The tray across the front holds pens. England, c. 1890. Brass. 8" x 12", 7 1/2" tall. $250-350.

The motif of this footed inkstand is the oak tree. Branches form the handle and the loose covers on the inkwells are acorns and leaves. The two bulbous ribbed inkwells fit in a cavity in the base. There are pen channels on both sides. America, c. 1885. Cast iron. The marble base has been painted. 7" x 14". $400-500.

Marbleized brown and white inkwell. The dome shaped lid has a brass collar and is hinged. America, c. 1900. 3 3/4" in diameter, 3" tall. $150-250.

Polished coal with a clear glass loose cover in a swirl pattern. America, c. 1920. 3" in diameter, 2" tall. $65-95.

Ink bottle with a screw on lid is fitted into a bakelite shell with a loose cover. The ink was probably purchased in this container America, c. 1915. 2 1/4" in diameter, 1 1/2" tall. $50-75.

The beehive shaped copper cover fits over a glass ink bottle. A stylized green eagle decorates the front. America, c. 1925. 3" in diameter, 2 3/4" tall. $65-100.

Pottery ink pot made in England, c. 1880. $35-50.

Hand-crafted soap stone inkwell. Early America, c. 1820. 2 1/16" x 2 1/8", 1 1/2" tall. $75-125.

Can shaped with a hinged lid. Advertisement reads "CALIFORNIA FRUIT CANNERY ASSOCIATION." A peach is embossed on the front and below in a banner is "Yellow Free Peaches." Under the lid is a hallmark and "Largest Canners In The World, San Francisco, Cal." America, c. 1900. Brass. 2" in diameter, 2 1/2" tall. $200-350.

Pottery ink pot made in England, c. 1880. $35-50.

Round wooden container with a screw on top that holds a glass ink bottle. Written on the bottle cap and the wooden lid "Sheaffer's SKRIP The Successor To Ink." America, c. 1930. 2 1/2" in diameter, 3 3/4" tall. $100-150.

Hand-carved wooden inkwell in the form of a jockey's head. (His ivory eyes are missing.) The cap is hinged in the back. Has a spring-loaded catch. England, c. 1900. 3" x 3 3/4", 5" tall. $250-300.

"The Non Spill" ribbed glass inkwell with a black plastic lid was made for the United States Navy by Preferred Products. c. 1935. $35-50.

Walnut stand inlaid with mother-of-pearl flowers and trim. The lid on the inkwell is hinged. The pen rack holds a matching pen with a mother-of-pearl flower on the handle. England, c. 1900. 4 3/8" x 6 5/8", 2 1/8" tall. $300-450.

Japanese writing desk encased in a cloisonné frame with bronze handles and trim. There are two open compartments and three lidded compartments with Foo dog finials. In the center of the box is an earthenware ink bottle. c. 1890. Mahogany. 11 1/2" x 14 7/8", 3" tall. $650-850.

Wooden inkstand with two clear crystal inkwells and a pen wiper brush seated in brass frames. A pen rack and letter holder are in the back. Stands on four small bun feet. England, c. 1915. Oak with brass mounts. 6 1/4" x 10", 5 3/4" tall. $500-600.

English mahogany writing desk trimmed in black. The two crystal inkwells have hinged lids. There is a round compartment in the center with a loose lid to hold pen nibs. The channel across the top holds pens. The drawer with a brass knob is for stationery and desk accessories. c. 1890. $600-750.

A footed stand that has two crystal inkwells inside wooden pegs on the base. The faceted lids have a brass collar and are hinged. The rear mounted pen rack is made with pegs. Across the front is a long box with a loose lid that opens to expose a four compartment stamp box. American folk art. c. 1920. Wood. 5 1/4" x 8, 5 1/4" tall. $250-350

Oak writing box with two glass inkwells with hinged lids. A channel across the front is for pens. The drawer below holds paper and other writing paraphernalia. The loop handle and hinged lids are silverplate. England, c. 1890. 6 1/2" x 11 1/4", 4 3/4" tall. $400-500.

English oak writing box with silverplated mounts. A pair of crystal inkwells with hinged lids are on each side. The drawer opens to store stationery and writing accessories. England, c. 1875. 9 1/4" x 13 1/2", 8 1/4" tall. $700-800.

English oak writing box with a brass hoop handle. Fitted with two crystal inkwells with brass hinged lids. There is a pen channel on the front and back. The drawer in front has an ivory pull knob and holds paper and other writing paraphernalia. Decorated with a beaded trim. c. 1890. 7 1/4" x 11 1/4", 6" tall. $400-600.

Mahoghany inkstand with two clear crystal inkwells. The faceted lids have brass hinged collars. A beveled mirror is in the back and a pen tray across the front. England, c. 1900. $300-500.

Wooden stand with an octagonal shaped crystal inkwell. The wooden cover is attached to a pole that swivels between two upright posts. To open the inkwell, the post attached to the lid is pushed down to lift the lid. England, c. 1900. Pine. 5" in diameter, 3 1/4" tall. $150-200.

Walnuts sitting on a large leaf with red berries nestled among them. The top leaf and nut form a hinged lid. c. 1880. Vienna bronze. 2 1/2" x 4 1/4", 4 1/4" tall. $300-400.

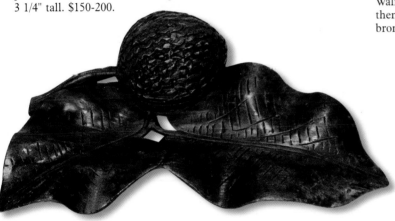

A walnut is resting on two large leaves. The nut is hinged in the back. Hand carved in Germany, c. 1905. 4" x 8", 2 3/8" tall. $150-250.

American folk art hand made from burl wood. The lid slides sideways to open. A glass insert is inside. c. 1920. 3 1/2" x 3 1/2", 2 1/4" tall. $60-95.

Twigs and oak leaves form the base. The removable lid is covered with acorns and leaves. Germany, c. 1880. Carved walnut. 7" x 7 1/2", 3" tall. $200-300.

Hand carved wooden hat with a feather in the hatband. The crown is hinged. Switzerland, c. 1900. 3 5/8" x 4", 2 1/4" tall. $150-250.

A crystal inkwell sits in a wooden base. The loose cover has a carved flower on top. America, c. 1925. 4" x 4", 3" tall. $75-100.

Art Deco stand with two inkwells that have hinged lids. The letter rack in back is constructed with leaves. Austria, c. 1920. Painted tin. 5 1/2" x 7", 3 1/2" tall. $100-150.

Matching mahogany desk boxes: one is an inkwell; the other is a pen holder. England, c. 1900. 2 1/2" x 2 3/4", 5 1/2" tall. $350-450 pair.

The roof is hinged on the wooden hut and opens to reveal a glass inkwell inside. Bavaria, c. 1890. 3 1/4" x 3 1/4", 3" tall. $75-125.

Hand carved wood in the shape of a log cabin. The roof is hinged. Germany, c. 1880. 3 1/2" x 6". $125-195.

Two wooden cabins sitting on a leaf. One holds an inkwell and the other is a stamp box. Marked "C. Ryon." Carved walnut. Black Forest, Germany, c. 1910. 4 3/4" x 12". $150-195.

Dark brown wooden hut. The top half is hinged and opens to reveal an inkwell inside. Germany, c. 1890. 3 1/4" x 3 1/4", 3" tall. $75-125.

Hand carved leaf with two glass inkwells that have leaf shaped lids. The tray across the front holds pens. The box in the middle has a hinged lid with pair of birds on top. Germany, c. 1880. Walnut. 6" x 12 1/2", 4" tall. $275-375.

Souvenir from "Marquette, Mich." Carved from a rough piece of wood complete with bark. A glass ink bottle is encased in the top and two tubes for pens are on the front. American folk art. c. 1920s. 3 1/2" x 5 1/2", 2 3/4" tall. $45-65.

Hand carved log cabin with a hinged roof. Written on the roof is "O. Olsern." Germany, c. 1890. 2 3/4", x 3 1/4", 3 1/4" tall. $125-200.

Hand carved in the shape of an old worn out shoe. The top of the shoe is hinged. Denmark, c. 1900. Wood. 2 1/4" x 5", 3" tall. $125-175.

Above: Treen ware beehive on a black base. The removable beehive is the cover for a glass ink bottle encased inside. England, c. 1860. 4 7/8" in diameter, 3 1/2" tall. $150-250.

Right: Carved treen ware with a pewter insert. There are three quill holes on the top. c. 1820. 2 3/8" in diameter, 1 3/4" tall. $150-250.

A silver bow decorates the top, and a silver swag with bows decorates the front of this round inkwell. England, c. 1900. Walnut. 3 3/4" in diameter, 2 1/4" tall. $125-175.

Coal bucket that has a hinged lid with a handle on top. England, c. 1900. Pine. 1 7/8" x 3", 2 3/4" tall. $65-100.

Round wooden inkstand with six quill holes. There are two recessed glass bottles: one is larger than the other. This particular inkwell was found in California gold mining country. Manufactured by S. Silliman and Company of Chester, Connecticut. America, c. 1850. 5" in diameter, 3 1/4" tall. $100-200.

Footed ball shaped inkwell with a matching lid. The interior is weighted with plaster and there is a long glass insert fitted inside. Marked on the bottom "Oriental." Bronze. 2" in diameter, 2 1/2" tall. $100-175.

Draftsman's inkstand. A stylus is attached under the lid. Marked "Philadelphia" on the stand, and on bottom of the bottle "Higgins Ink." America, c. 1920. Iron. 2 1/2" x 5 1/2", 3 1/2" tall. $150-275.

This inkwell has the appearance of a mousetrap. In order to open the ink bottle the wire frame must be snapped off of the top. The bottle has a stylus attached under the lid. "Higgins" is inscribed on the base of the inkstand and "Higgins Drawing Ink" on the bottle. America, c. 1920. 4" x 6 1/4". $150-250.

Inkwell attached to a clip that fits over a notepad. Screw on cap. Marked "Chumbacher 833 Japan." c. 1935. Steel. 2" in diameter. $50-75.

Draftsman's inkstand. The lever bar on the back opens the cover. Written on the top of the base "LIETZGEN" and on the steel clamp "Pat. 1772610." 1930. 2 3/8" x 4", 3" tall. $150-275.

Irish bogwood in the shape of a three-legged kettle is covered with shamrocks. The loose cover with a ball finial has "INK" written on the top. c. 1890. 1 3/4" in diameter, 2 1/4" tall. $75-125.

Traveler's Inkwells

Top hat and an ivory handled cane. When the crown of the hat is removed, the inkwell is revealed. The cane has a pen on one end and pencil on the other. 5 1/2" long, 1 3/4" tall. $400-600.

The pieces of the top hat and cane when disassembled.

Hat and umbrella. The top of the hat is removed to expose an inkwell with a screw on lid. The umbrella comes apart and is a pen and pencil. Written on the hat "Bad Ems." Painted aluminum. 4 5/8" long. $350-500.

The top of the hat on the fisherman's head is hinged and opens to reveal an inkwell with a spring-loaded lid. Carved and painted wood. England, c. 1900. 2 3/4" x 3", 3 3/4" tall. $250-300.

A square wooden box that is constructed for travel. The lid snaps open to expose a glass ink bottle fitted inside. There are four round holes around the bottle to store extra pen nibs. Made in England, c. 1860. Brass fittings. 1 1/2" x 1 3/4", 2" tall. $300-400.

Above: This wooden traveler is called a "Penner" and consists of three parts: one section holds the ink bottle, one the pen nibs, and on the top section is a seal. England, c. 1850. 1" in diameter, 4" tall. $300-400.

Right: Closed view of the three-part penner.

Above: Open view of the horn showing the inkwell on the right and a sander on the left.

Left: This "Penner" also consists of three sections. When unscrewed, the bottom piece is a sander and the middle piece is an inkwell. England, c. 1840. Made from a horn. 1 5/8" in diameter, 6 1/8" tall. $400-500.

Two-pocket set, an inkwell and matching matchbox. The inkwell has a hinged lid that snaps tight over a glass ink bottle inside. The match container has two hinged lids; the inner lid has a ribbed striking plate for the matches. The top lid closes the box. England, c. 1880. Brass. 1 1/4" x 2 3/8", 2 1/8" tall. $150-200 each.

The legs on this traveler were made to fit onto the arm of a chair; when not in use, they fold up into the hollow bottom. Winged cherubs are on both sides and the top of the lid. On the bottom is a Dominick & Haff hallmark. Inscribed "Mermod & Jaccard & Co. Sterling. Patent. 24." Mermod & Jaccard were jewelry dealers in St. Louis, Missouri. America, c. 1875. 3" in diameter. $400-500.

A man's hat made of stiffened fabric. The spring release button is in front under the grosgrain ribbon hatband. The crown is hinged and opens to reveal the inkwell. 3 1/2" x 4", 1 3/4" tall. $250-400.

A small pressed glass traveler with a nickel-plated hinged lid that is attached to a four-ply felt pen wiper. Cloth pen wipers were often made as gifts, sometimes embroidered with flowers, names, and dates. America, c. 1900. Pen wiper: 4" x 4". $75-125.

Tiny traveler with a sliding cover that locks down with a wing nut. England, c. 1840. Pewter. 3/4" x 1 1/8", 1 1/4" tall. $400-500.

Above: Wooden barrel with a glass bottle inside and a screw on cover. Made by the Silliman Company, Chester, Connecticut. America, c. 1855. 1 1/2" x 2 1/4". $125-150.

Right: Tin helmet with a leather strap across the front. The top is hinged and opens to expose the inkwell inside. France, c. 1916. 3" x 3 1/4", 1 3/4" tall. $200-250.

A travel inkwell in the form of a thermometer case.

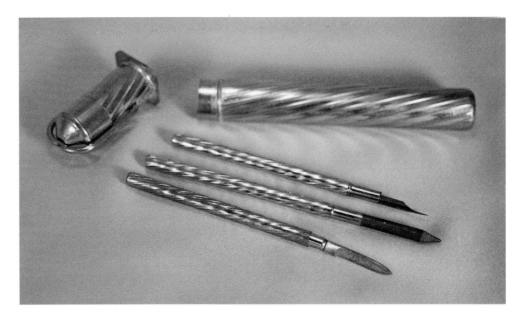

Open view of the thermometer case. A safety bar snaps down and the top unscrews to expose the inkwell inside. The bottom part of the case holds a pen, ink eraser, and a pencil. England, c. 1880. $200-300.

Marked "Ransome's Patent. De La Rue & 02." When the lid is opened, the ink bottle can be moved to an upright position. France, c. 1885. Wood and nickel silver. 2" x 2 3/8". $225-300.

Open view showing the encased ink bottle in an upright position.

Inkwell encased in a tin hatbox with a brass lid, straps, and handle. This should have been covered with leather. 1 7/8" in diameter, 2 1/8" tall. $75-125.

Inkwell with a square glass bottom and a brass top. The hinged lid is secured with a folding clasp that unscrews to open. c. 1850. 1 1/2" x 1 1/2", 1 3/8" tall. $150-250.

Traveler with a round glass bottom and a brass top. The hinged lid has a folding clasp that unscrews to open. c. 1850. 1 3/4" in diameter, 3" tall. $150-250.

Two inkwells. Left: English with screw down lid. Right: French with "Ransomes (crown) Patent. De La Rul & Co." on the front. c. 1900. Push to open. 1 1/2" x 2". Base metal. $125-195 each.

A novelty in the shape of a violin case. When the case is opened, an inkwell and pen wiper are exposed. The outer case has a push button lock and the inkwell inside has a spring-loaded catch. This double locking device prevents accidental spills. England, 1880. Brown leather over brass. $250-350.

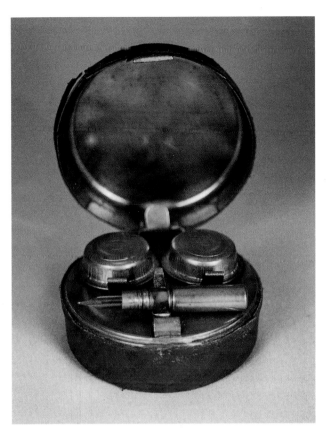

Open view of a spring release traveler. The two hinged inner lids snap open to reveal glass ink bottles. A pen is in a cylinder that is held in a clip in front; it unscrews for use. Inscribed in gold on the top "INK." The brass case is covered with red leather. England, c. 1875. 2 3/8" in diameter, 2 3/8" tall. $300-450.

Open view of a brown leather traveler. The two hinged inner lids open to expose ink bottles inside. A pen wiper brush is in front. "INK" is inscribed in gold on the top. England, c. 1875. 2" x 3", 1 1/2" tall. $350-500

The brass box is covered with red leather and is inscribed "INK" on the top in gold letters. The outer case is hinged and opens to reveal a beautifully engraved interior with two lidded inkwells and a pen wiper brush. England, c. 1870. $300-400.

Open view that shows two inkwells with hinged lids, a pen that snaps into a holder, and a hinged spring-loaded cylinder for holding a stick of sealing wax or candle. The wax is pushed down into the cylinder and is then folded over into a depression, so the lid will close. Brass with a leather cover. Inscribed in gold letters on the top "INK." The brass inside is covered with an engraved design. England, c. 1870. $400-550.

Souvenir in the form of a football. Inscribed in gold on the top "Montreal." When the hinged outer lid is opened, it exposes a second hinged lid with a spring lock that covers the glass ink bottle. England, c. 1880. Brass covered with brown leather. $275-350.

Leather over brass. "INK" is stamped on the lid in gold letters. The spring lock button is on the front under the leather. The glass ink bottle inside is covered with a spring loaded lid. England, c. 1885. $150-250.

A container marked "Waterman Ink." Nickel silver case with a glass bottle inside. The ink bottle has a stopper and a screw on cap. America, c. 1900. 1 1/2" in diameter, 3 3/8" tall. $45-75.

Covered with red leather. "INK" is printed in gold on the top. There is a button under the leather on the lower front to open the lid. Encased inside is a glass ink bottle with a spring lock cover. England, c. 1890. 1 7/8" in diameter, 1 5/8" tall. $150-250.

Souvenier with "Jerusalem" written on top of the hinged lid. The glass inkwell encased inside has a cover secured with a spring lock. Olivewood. $95-150.

Leather covered metal in the shape of a mailbox. Stamped on the front "U. S. Mail." The two lids are secured with spring locks. The button that opens the lid is under the leather in front. America, c. 1880. 1 1/8" x 2 1/4", 2 7/8" tall. $200-300.

Back of the mailbox showing the postal rates. Stamped in red is "Drop Letters Here, Letters U. S. Mail. Postal District M."

Glass with a screw on silver top. There is an inner metal stopper with a ring pull. This type would fit in a lap desk. England, c. 1880. 2" x 2", 1 1/2" tall. $75-95.

Black leather covered with a tooled flower on the top. There is a button on the front that opens the lid to expose a second spring loaded lid that covers the glass inkwell. The double locks insure a spill-proof ink container. Engraved inside under the lid is a double-headed eagle surmounted by a crown with an anchor that has a fish wrapped around it. Engraved "KKA-PRIV." Nickel-plated brass. 1 3/4" in diameter, 1 5/8" tall. $175-250.

A glass ink bottle is encased in a wooden cylinder with a screw on lid. Silliman & Company, Chester, Connecticut. America, c. 1865. 1 1/8 in diameter, 2 1/4" tall. $45-75.

This plain wooden traveler with a screw on cap was issued to American Civil War soldiers, 1861-1865. Silliman & Company, Chester, Connecticut. America. 1 1/8" in diameter, 2 1/8" tall. $50-100.

The lid is hinged and opens to expose a glass ink bottle inside. Embossed on the top half is "Johann Hoff." Germany, c. 1915. Brass. $300-400.

A green glass ink bottle is inside the aluminum container. A cork stopper and a screw on lid protect the ink from spilling while being transported. England, c. 1900. $125-150.

A jug with a hinged top opens to expose a glass ink bottle. Marked "Selters" and "Deposé." France. $250-350.

Japanese scribes pen holder with an attached inkwell that has a hinged lid. The inkwell holds dry ink that is mixed with saliva to liquefy the ink. The flower shaped container is called a Suiteki. The carved ivory rooster on the right is a netsuke. Bronze. 7/8" x 7". $195-250 (inkwell only).

North African scribe's pen holder with an attached inkwell that has a hinged lid. Brass. $200-300.

Oriental scribe's case. The footed inkwell has a hinged lid. The long pen or brush holder has vines and flowers decorating the top. Bronze. 2 3/8" wide x 10 3/4" long. $200-300.

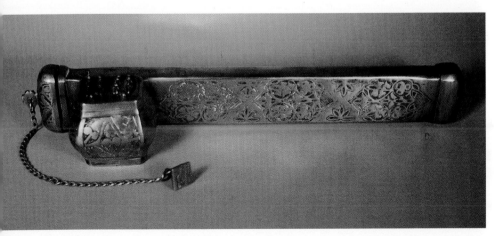

Oriental scribe's writing case to be carried in a sash around the waist. The long compartment holds pens or brushes. The small box on the side contains two small inkwells: one for black ink, the other for red. There is a seal on the end of the attached chain. c. 1880. Beautifully incised design on brass. Over 11" long (larger than most). $300-400.

References

Books

Badders, Veldon. *Collectors Guide To Inkwells*. Paducah, Kentucky: Collector Books, 1995.

Covill, William E., Jr. *Ink Bottles and Inkwells*. Taunton, Massachusetts: William S. Sullwold, 1971.

Godden, Geoffrey A., F.R.S.A. *Encyclopaedia of British Pottery and Porcelain Marks*. New York: Bonanza Books. MCMLXV.

Goodman, Joel. *A Better Pictorial Book of Inkwells and Accessories*: Joel Goodman. 1980.

Kovel, Ralph and Terry. *New Dictionary of Marks*. New York: Crown Publishers, Inc., 1986.

McGraw, Vincent D. *Mcgraw's Book of Antique Inkwells*, vol. I. Vincent D. McGraw, 1972.

McNerney, Kathryn. *Antique Iron*. Paducah, Kentucky: Collector Books, 1984.

Nickell, Joe. *Pen, Ink, & Evidence*. Lexington, Kentucky: The University Press of Kentucky, 1990

Rainwater, Dorothy T. and H. Ivan. *American Silverplate*. Nashville, Tennessee: Thomas Nelson Inc. and Hanover, Pennsylvania: Everybodys Press., 1968.

Rainwater, Dorothy T. *Encyclopedia of American Silver Manufacturers*. Third edition revised. West Chester, Pennsylvania: Schiffer Publishing Ltd., 1986.

Rivera, Betty and Ted. *Inkstands and Inkwells A Collectors Guide*. New York: Crown Publishers, Inc., 1973.

Publications

Patents and Their Time Frames. San Jose, California: United States Patent Office.
The Stained Finger. Minneapolis, Minnesota: The Society of Inkwell Collectors. Vincent McGraw.

Catalogs

Marshal Field & Co. 1896 Catalogue Reproduction. Northfield, Illinois: Gun Digest Publishing Company, 1970.

Pairpoint Manufacturing Co., 1894 Catalogue Reproduction. Washington Mills, New York: The Gilded Age Press, 1979.

Sears, Roebuck Catalogue 1897, 100th Anniversary Edition. Philadelphia, Pennsylvania: Chelsea House Publishers, 1993.

Sears, Roebuck Catalogue 1908 Reproduction. Northfield, Illinois: Digest Books, Inc.

Index

Acorns and oak leaves, walnut, 269
Agate, 188
Alligator and stump, 64
Amber cut crystal, 233
American folk art, wood, 267, 269, 272
Angelus Call to Prayer, 87
Antelope, 57, 58
Arab on Oriental rug, 86
Arab with pipe, 87
Armistice train, 109
Art Deco, 204
Art Deco double inkwell, 186
Art Deco hexagonal, 115
Art Deco rolled brass, 213
Art Deco single, 114, 169, 170
Art Deco single stand, 213
Art Deco, tall lady, 70
Art Deco, black glass, 259
Art Deco, square, black glass lid, 259
Art Nouveau double, 206
Art Nouveau, cherub on seashell, 243
Art Nouveau, Egyptian style, 214
Art Nouveau, floral, 206
Art Nouveau, portrait on sterling lid, 241
Art Nouveau, single 170, 185, 190
Art, science, and engineering, 107
Arts and Crafts, 113, 114
Athenian, made for Marshal Fields, 113
Austrian faience, triangular, 125
B. R. S. Limoges, porcelain, 124
Baby bottom side up, 261
Baccarat, Gorham sterling lid, 240
Baccarat, green to clear, 234
Bakelite with lizard, 111
Baker chocolate lady, 72
Ball pyramid, 250
Barometric, 166
Basket with roses on handle, 127
Basket, brass wire, 99
Bastard Park, Made in France, 124
Battleship, 97
Bears, 41, 42, 43
Beetle with jeweled back, 64
Bell, blue and white porcelain, 120
Bicyclist, 177
Birds on a leaf, walnut, 271
Birds, 22, 25
Bitch with five puppies, 41
Black boy in floppy hat, 81
Black glass, 225
Black glass, square, 112
Black lacquered, double, 111
Black man in white robe, 89
Black man playing a flute, 89
Black man's head, 89
Black ware, 110
Black, deep recessed saucer, 129
Blond man with letter, 151
Bloor Derby, porcelain, 129
Blue and gold ceramic, 121
Blue and White flying crane, 119
Blue and white spatter, 112
Blue and white with windmill, 126
Blue and white, sliding lid, 144
Blue flowers, saucer base, 112
Boar's head, 63
Boat, mother of pearl, 100
Bohemian glass, 176
Bohemian glass, red to clear, 228
Boulle work writing desk, 191

Boy beside a well, 84
Boy feeding a dog, 31
Boy hunter, inkwell, match holder, 84
Boy in clown hat, 75
Boy pushing wheelbarrow, 72
Boy with rabbit by his foot, 85
Boy with scythe, 84
Boy with spilt ink, 75
Boy wrestling with a goose, 81
Boy's head with a cap, 83
Boys with bird nest, 148
Bradley & Hubbard, brass, 203
Bradley & Hubbard, single brass, 213
Brass & crystal, 249
Brass and milk glass, 202
Brass plated, 197
Brass single, 208
Brass stand, 202
Brass urn, footed, 203
Brass with blue & white wells, 199
Brass with crystal, 195, 198, 199, 201
Brass with enamel flower, 198
Brass with glass well, 201
Brass, double, 169, 209
Brass, double, England, 212
Brass, double, Germany, 212
Brass, screw on lid, 252
Brass, single on pedestal, 190
Brass, single with tray, 215
Brass, single, 217, 220
Brass, single, France, 211
Brass, single, square, 210
Brass, square with handles, 207
Brilliant blue & clear crystal, 233
Brilliant blue crystal, 229, 230, 231, 232
Bristol blue, Sheffield lid, 223
Bronze doré with candle holder, 196
Bronze on black marble, 200
Bronze overlay on wood box, 119
Bronze with swags, 203
Bronze with enameled portrait, 196
Bronze, cherries & pods, 206
Bronze, footed urn on pedestal, 214
Bronze, single, 190, 205
Buffalo, 55, 56
Bulbous crystal with sterling lid, 242
Bulbous shaped, striped, 138
Burgundy red and gold porcelain, 122
Calendar, 172
Camels, 43, 45, 46
Cameo glass, Thomas Webb, 224
Can shape, California fruit, 264
Canada, souvenir, 110
Candle holder and inkwell, 128
Capodimonte, 131, 132, 143
Carlton ware, oriental scenes, 123
Cart with wheels, wood, 95
Carter engraved on front, 241
Casket with embossed flowers, 117
Casket with oriental garden inside, 117
Cats, 29, 30, 31
Cavalier with feather in hat, 80
Celedon hexagonal, 131
Ceramic with red berries, 135
Chains, 188
Champlevé 121, 122
Cherub and fairy, 73
Cherub on pedestal, 73
Cherub with wings, 73
Chess players, 149

Chestnuts on leaf, 213
Child on bed, 74
Child's face in bonnet, 73
Chintz, 137
Chrysanthemums in a circle, 114
Clear crystal, blue enamel, 245
Clear crystal, brass lid, 243
Clear crystal, sterling lid, 240
Clear glass, black glass funnel, 259
Cleopatra's head on tray, 69
Clock, sterling, 238
Cloisonné, 121, 126, 209
Cloisonné writing desk, 266
Clown, 74, 75, 79
Coal bucket, 273
Coal, folk art, 262, 263
Cobalt blue with gold, 133
Continental silver, double, 180
Controlled bubble set, 246
Copper cover over glass bottle, 264
Copper with pen rack, India, 116
Cornucopia, Mottahedeh, Italy, 133
Cornwall fairy with mushroom, 76
Cottage with thatched roof, 86
Counting house, 171
Couple at harpsichord, 148
Coventry Corners, 110
Crabs, 64, 65, 66
Cranberry glass on lily pad, 225
Crescent shape, crown lid, 242
Croquet mallet-tennis racket, 190
Crystal brick house, 247
Crystal in wooden base, 270
Crystal with brass domed lid, 250
Crystal, gadroon decorations, 242
Crystal, lid with crossed flags, 252
Curling stones, 102
Cut crystal, faceted lid, 254, 256
Cut crystal, mushroom lid, 244, 252, 255
Cut crystal, pyramid steps, 244
Cut crystal, sterling lid, 235
Dachshund puppies on pen tray, 40
Daisy & button vaseline glass, 234
Daisy & Button, blue pressed glass, 228
Dark blue porcelain, France, 126
Deep dish with quill holes, 137
Deer heads on bee hive well, 54
Deer hoof, 57
Deer reclining, 52
Delft, Holland, 133
Delft-like with cherubs, 151
Denmark, B. S. Grondal, 144
Derby silver plate, single, 218
Desk boxes, ink and pen holder, 270
Diamond shaped crystal, tiny, 255
Diamond shaped cut crystal, 254
Diogenes holding a lantern, 88
Doe with fawn, 47
Dogs 31, 32, 33, 34, 35, 36, 37, 38, 39, 40, 41
Donkey, 47
Doré bronze, green wells, 187
Double brass with stamp box, 201
Double porcelain, handle, 139
Double with candlestick, 130
Double with sponge cup, 163
Double with stamp box, 162, 164
Double, Bakelite lids, 257
Double, brass, 162
Double, cast iron, 161, 162, 167, 168, 169
Double, leaves form pen rack, 270
Doulton, Lambeth, 141

Doves, 21, 27
Draftsman's inkstand, 274
Dresden, bowl on saucer, 136
Dresden, porcelain & sterling, 223
Dresser drawers, 140
Drummer with drums, 147
Duck, child's inkwell, 21
Dutch girl, 71
E. Gallé Nancy, 143
Eagles, 25, 26, 27
Egyptian Pharaoh on bowl, 69
Electric lamp held by boy, 262
Elephant foot, 185
Elephant, 60, 61, 62
Emerald green glass, bronze, 222
Erphila Ink Girls, 145
Excelsior Inkstand, brass top, 248
Expo Colonial 1931, souvenir, 109
Faceted crystal, black lid, 249
Faience, France, 132, 134, 176
Faience, Italian, 133, 142
Ferryboat, 101
Fire hydrant, Philadelphia, 105
Firenze, Italy, 171
Fish on sides, 171
Fish, waffle pattern, 243
Flowers on frosted medallions, 239
Footed ball shaped, 273
Footed crystal, Gorham sterling lid, 242
Footed with handles, single, 218
Footed with two wells, 198
Four compartment, iron, 217
Four crossed muskets, Tufts, 96
Four inkwell stand, 261
Foxes, 55
France, souvenir, 108
French Foreign Legion, 79
Friar with a cup and keys, 81
G.D.A. France, porcelain, 126
Geese and peacock, 28
Gentleman sitting on a stile, 77
George Washington on horse, 147
Germany, double with stamp box, 208
Girl in bonnet, 72
Girl paddling a boat, 70
Girl sitting with crossed legs, 69
Glass and brass, 260
Glass boat, opaque, 111
Glass casket, 254
Glass dog, 34
Glass with attached funnel, 101
Glass with Bakelite sliding lid, 258
Glass with nickel plated collar, 246
Glass with two dip holes, 254
Glazed porcelain, single, 135
Goat pulling a cart, 46
Goat's head with bells, 46
Goats with herder, 46
Golfer, 178
Gondola shaped porcelain with seal, 129
Gorham sterling, bell shaped, 182
Gouda pottery, Holland, 125
Grand piano, 103
Grapes on leaf, 140
Grapes, brass plated iron, 197
Grapes, cast iron, 200
Gravity fed, glass, 251
Green art pottery, speckled, 123
Green glass with jeweled lid, 235
Green porcelain, double, 148
Green to clear glass, 235
Gutta-percha, 193
Hall's toffee box, 106

Hand blown with stopper, 224
Hat with hatband, wood, 270
Head of woman in cap, 93
Heart shaped with roses, 125, 145
Heater, 104
Heavy blown glass with dip hole, 259
Heintz Art Metal, silver overlay, 113
Helmets, 97, 98, 99
Hexagonal art glass, clear, 253
Hexagonal crystal, 244
Hexagonal cut crystal, 253
Hibiscus design tray, crystal well, 261
Higgins inkstand, mouse trap, 274
Holly, CFH over GDM, 132
Horns on round base, 104
Horns with pen holder on top, 103
Horse and cowboy, 50
Horse hoof, pet pony, 51
Horse tether, 100
Horseman with banner, Paris, 108
Horses, 48, 49, 50, 51, 52
Horseshoes, 49, 50, 51
Horseshoe, green glass, 234
Hunter with gun, porcelain, 84
Hut by millstream, 85
Hut with match holder, 86
Hut with thatched roof, 219
Hut woven brass, 86
I. Magin & Co. green crickets, 128
Imp or demon, 75, 76
Indian head, 82, 83
Indian woman in robe and turban, 71
Ink bottle, screw on lid, 263
Inkhorn, 103
Inkstand with candelabra, brass, 218
Inkwell & sander, tan, 141
Inkwell from inkstand, 255
Iridescent blown glass, 256
Irish bog wood, 274
Irish pewter, 179
Iron, double, 215
Ivory pyramid of snooker balls, 124
Jacob Petit, floral with gold, 150
Jeweled, 161, 187, 197
John Bull with chicken, 78
K. & O. Metal Novelties Co., 116
Kaiser Bill in helmet, 82
Kettle with three legs, 118
Kidney shaped mirrored tray, 130
King Rex, Mardi Gras, 91
Kiss or Kisc, square porcelain, 125
Kittens, 28, 29
Kiwi, 24
Lalique style frosted glass, 236
Lantern, 101
Large oriental with foo dog, 207
Large stand, rotary rack, 262
Laughing man framed, 95
Leaf with crystal, 200
Light bulb, 101
Lily pad with amethyst glass, 214
Limoges, France, turquoise, 136
Limoges, horse riders, 138
Lions, 58, 59, 60
Lizards, 64
Loetz style, 220, 221, 222, 227
Lotus leaves, 210
Lusterware, 141, 142
Mahogany writing desk, 267, 268
Majestic stag, 53

Mandolin on sheet music, 102
Marble stand, 194
Marbleized brown & white, 263
Mask on lid, 94
Match box with inkwell, 159, 160
Matthew's-Marves-Lycas, Johannes, 115
Men, 77, 78, 79, 80, 87, 88, 90, 91, 92, 93
Mermaids, 67
Metal, double, 170
Mice with cheese and crackers, 62
Middle Eastern scribes inkwell, 115
Milk glass with brass top, 145
Millefiori, paper weight inkwell, 240
Monks, 80, 90
Moroccan wearing a red fez, 89
Mosaic, tiny, 121
Mount Washington, 226
Mr. & Mrs. Carter Inx, 72
Mule pulling a cart, 47
Mythological creatures on globe, 198
Napoleon, 92
Navy, Preferred Products, 265
Nippon, 120, 134
Note pad clip, 274
Notre Dame, souvenir, 107
Oak tree branches, acorns, 263
Oak writing box, 267, 268
Occupied Japan, 134
One horn with well, 104
Onyx base, open work back, 217
Opalescent glass, pressed brass, 260
Open work brass base, crystal, 261
Orange porcelain, brass top, 112
Oriental casket, 116
Oriental design, sliding lid, 118
Oriental garden scene on black porcelain, 127
Oriental style with water dragon, 119
Oriental teapot with handle, 118
Ornate stand with face, double, 220
Oval stand, ring handle, 207
Overlay, grapes on glass, 219
Owl, 16, 17, 18, 19, 20
Pageboy with a letter, 85
Paneled bulbous shaped glass, 247
Paneled star cut, mushroom lid, 253
Pansies, 195
Paperweight, red to clear, 228
Paperweight-partners inkwell, 224, 225
Papier-mâché, 192, 193
Parakeet, 24
Paris, France, 108
Paris, souvenir, 109
Parrot, 22, 24
Partners inkwells, 178
Partners, double brass, 205

Partners, double bronze, 216
Partners, single brass, 206
Pedestal with winged handles, 196
Pegasus, swans, double brass, 216
Peking enamel, hexagonal shape, 119
Perry & Co. gravitating, 200
Pheasants, 28
Phrenology head, 93
Pietra Dura Italian mosaic, 189
Pinocchio, wooden, 91
Pistol with shot and casings, 96
Plain Brass, single, 218
Porcelain leaves, gold trim, 150
Porcelain with brass top, France, 127
Porcelain with cups on the side, 130
Porcelain, blue with white, 146
Porcelain, looks like pottery, 135
Porcelain, swags and red roses, 136
Portrait in plated lid, glass well, 239
Posy, 176
Pottery ink pot, 264
Pressed glass with tray, 239
Pressed glass, brass lid, 246
Pressed glass, mushroom lid, 253
Pump inkwell, 173, 174, 175
Puppy, 35
Pyramid, clear crystal, 251
Pyramid, vaseline glass, 251
Quail with chicks, 24
Quimper, faience, 143
Rabbits and carrots, 63
Railroad inkwell, glass, 245
Ram on hoofed feet, 47
Rat with parsnip, 63
Red ware, three compartment, 110
Red, white, and blue enameled, 120
Repoussé lid on crystal, 185
Ribbed blue glass, 223
Richard Wagner, 91
Rococo, iron with glass, 172
Roll top, 161, 163, 169
Roman soldier, 78
Roosters, 20, 21
Rose with stem, 204
Round blown glass, 249
Round crystal, sterling lid, 241
Round glass paperweight, 249
Round glass with metal lid, 249
Royal Crown Derby, 144
Royal Doulton, 141
Roycroft, arts and crafts, 112
Ruby glass, 228
Ruby glass, white overlay, 227
Saddle and horse blanket, 95
Sailing ship, The Victory, 99
Samson, octagonal inkwell, 123
Sarreguemines, red sofa, 135
Satin glass, 237
Saucer, turned up edge, 197
Scale weight, 100

Scale, 171
Scalloped edged porcelain, 140
Sea creature with shell, 188
Seals in ocean waves, 62
Seashells & candle holder, 183
Seashells, bronze, 66
Sengbusch with Bakelite lid, 258
Sheaffer's SKRIP, wood & glass, 265
Sheffield double, 187
Sheffield gallery tray, 183
Shield shaped with sterling lid, 245
Ship, mother-of-pearl, 100
Shoe, Rockingham, 77
Shoe with hole in toe, 272
Shoe, glass, 260
Shreve & Co. Sugar bowl, 179
Shreve & Co., San Francisco, 247
Silliman & Co., wood, round, 273
Silver plate casket, England, 179
Silver with moose head, 181
Silverplate, 180, 181, 186, 194
Single castellated top, 176
Single with scrolls & leaves, 210
Single, cast iron, 160, 163, 164, 165, 166, 167, 168
Six sided crystal, sterling lid, 248
Skulls, 94
Snail type, double, 153, 154, 155, 156, 157, 159
Snail type, single, 151, 152, 153, 154, 155, 156, 157, 158
Snail type, triple, 152, 153
Snakes & mother-of-pearl, 189
Soap stone, hand carved, 264
Souvenir Catalina Island, 105
Souvenir in shape of boat, 106
Souvenir in shape of fish, 106
Souvenir, two, carved bone, 107
Spain, hexagonal, 138
Spaniel and quails, 40
Spaniel on Rococo stand, 40
Spanish soldier, 92
Square crystal, brass cover, 247, 252, 256
Square crystal, silver lid, 248, 249
Square cut crystal, domed lid, 250
St. George, Real Bronze, 210
Staffordshire, 144, 145, 146
Stag, 52, 53, 54, 55, 184
Stag's head, cranberry wells, 226
Stained glass double stand, 236
Stand with cherub seal, 131
Stand with four compartments, 208
Standish, three compartment, 188
Star shaped brass with glass, 211
Sterling and carrara marble, 182
Sterling and enamel, single, 181
Sterling overlay, 177, 178
Sterling overlay, green crystal, 234
Sterling silver and glass, 180
Sterling stand, crystal well, 185

Sterling with two crystal wells, 182
Sterling, single with iris, 184
Sterling, stamp box and roller, 179
Stork, 22
Stove, footed, 105
Swan house, 21
Sweden, silverplate, 183
Swirl glass with sterling lid, 243
Swirl pattern glass, large, 241
Tatum, double glass, 257
Tea kettle, 136, 137
Tea kettle, cranberry glass, 227
Tea kettle, green glass, 235
Thermometer, 166
Thermometer, single crystal well, 236
Three compartments, handle, 150
Three people with baby, 149
Three wheel bath cart, 102
Tiffany single, 189
Tiffany, 204, 205
Tiffany, favrile glass, 223
Tiffany, sterling, 182
Tiger-eye, 188
Tiny cap, 196
Tiny cut crystal, sterling lid, 248
Tiny pressed glass, swirl, 251, 252, 255
Train, mahogany, 95
Traveler's inkwells
  Aluminum, 284
  Black leather with flower, 283
  D. & H. sterling silver, 277
  Double with pen holder, 281
  Double with wax or candle holder, 281
  Double with wiper brush, 281
  Double, INK on the lid, 281
  Fisherman, 276
  Football, Montreal, 282
  Glass with screw on lid, 283
  Hat box, 280
  Hat with hatband, 278
  Japanese scribes, 285
  Johann Hoff, 284
  Mail box, 283
  North African scribes, 285
  One England, one France, 280
  Oriental scribes, 285
  Penner, horn, 277
  Penner, wooden, 276
  Pressed glass on wiper, 278
  Ransome's, 279
  Red leather, INK on the lid, 282
  Round glass with brass lid, 280
  Scribes writing case, 285
  Selters, France, 284
  Silliman, wood, 278, 284
  Single leather with INK on lid, 282

Souvenir, Jerusalem, 282
Square box, 276
Square glass with brass lid, 280
Thermometer case, 279
Tin helmet, 278
Tiny pewter, 278
Top hat, 275
Two pocket, match and inkwell, 277
Violin case, 280
Waterman Ink, nickel silver, 282
Tree branches with crystal wells, 238
Treen ware beehive, 272
Treen ware, round, 272
Triangular cut crystal, 245
Triangular rolled brass, 115
Triple glass stand, 257
Triple inkwell with lids, 111
Turtle, 63
Urn on pedestal, brass, 214
Urn with marble well and base, 105
Urn, single on onyx base, 219
Walnut stand with mother of pearl, 266
Walnut wood with silver bows, 273
Walnuts and berries on leaf, 269
Walnuts on a leaf, 269
Walrus head, 62
Washerwoman, 146
Washington, D. C., 108
Watch stands, 237
Wave Crest, 224
Webster, E. G. & Son, square, 120
Wedgewood, 175
Wheat basket, porcelain, 137
White porcelain with bird, 139
White porcelain with green leaves, 134
White porcelain with turtle, 139
White porcelain, brass top with seal, 128
White porcelain, lake scene, 139
White with applied flowers, 132
White with pink flowers, 138
Winged cherub on stump, 80
Winged dragon with coiled tail, 79
Wolves tracking a man, 48
Woman, 66, 67, 68, 69, 70, 71
Wood with brass and crystal, 195
Wood with octagonal bottle, 269
Wood, brass, & glass stand, 266
Wooden cabins, Black Forest, 271, 272
Wooden hut, Black Forest, 271
World War I soldier, 96
Writing paper with quill, 170
Yellow goblet inside blown glass, 240
Yellow porcelain, brass base, 131
Zimmerman, 199